Content-Based Image Classification

T0341069

Content-Based Image Classification

Efficient Machine Learning Using Robust Feature Extraction Techniques

Rik Das

CRC Press
Taylor & Francis Group
Boca Raton London New York

CRC Press is an imprint of the
Taylor & Francis Group, an **informa** business

A CHAPMAN & HALL BOOK

MATLAB® and Simulink® are trademarks of The MathWorks, Inc. and are used with permission. The MathWorks does not warrant the accuracy of the text or exercises in this book. This book's use or discussion of MATLAB® and Simulink® software or related products does not constitute endorsement or sponsorship by The MathWorks of a particular pedagogical approach or particular use of the MATLAB® and Simulink® software.

First edition published 2021
by CRC Press
6000 Broken Sound Parkway NW, Suite 300, Boca Raton, FL 33487-2742

and by CRC Press
2 Park Square, Milton Park, Abingdon, Oxon, OX14 4RN

© 2021 Taylor & Francis Group, LLC

CRC Press is an imprint of Taylor & Francis Group, LLC

Reasonable efforts have been made to publish reliable data and information, but the author and publisher cannot assume responsibility for the validity of all materials or the consequences of their use. The authors and publishers have attempted to trace the copyright holders of all material reproduced in this publication and apologize to copyright holders if permission to publish in this form has not been obtained. If any copyright material has not been acknowledged please write and let us know so we may rectify in any future reprint.

Except as permitted under U.S. Copyright Law, no part of this book may be reprinted, reproduced, transmitted, or utilized in any form by any electronic, mechanical, or other means, now known or hereafter invented, including photocopying, microfilming, and recording, or in any information storage or retrieval system, without written permission from the publishers.

For permission to photocopy or use material electronically from this work, access www.copyright.com or contact the Copyright Clearance Center, Inc. (CCC), 222 Rosewood Drive, Danvers, MA 01923, 978-750-8400. For works that are not available on CCC please contact mpkbookspermissions@tandf.co.uk.

Trademark notice: Product or corporate names may be trademarks or registered trademarks and are used only for identification and explanation without intent to infringe.

Library of Congress Cataloging-in-Publication Data

Names: Das, Rik, 1978- author.
Title: Content-based image classification: efficient machine learning using robust feature extraction techniques / Rik Das.
Description: First edition. | Boca Raton: C&H\CRC Press, 2021. | Includes bibliographical references and index.
Identifiers: LCCN 2020030222 (print) | LCCN 2020030223 (ebook) | ISBN 9780367371609 (hardback) | ISBN 9780429352928 (ebook)
Subjects: LCSH: Optical pattern recognition. | Images, Photographic-- Classification. | Image analysis--Data processing. | Image data mining. | Machine learning.
Classification: LCC TA1650 .D376 2021 (print) | LCC TA1650 (ebook) | DDC 006.4/20285631--dc23
LC record available at https://lccn.loc.gov/2020030222
LC ebook record available at https://lccn.loc.gov/2020030223

Typeset in Palatino
by MPS Limited, Dehradun

I express my gratitude to the Almighty for bestowing love, affection and blessings

on me and showing me all possible routes leading toward successful completion

of this book.

I wish to dedicate this book to my father, Mr. Kamal Kumar Das; my mother, Mrs.

Malabika Das; my wife, Mrs. Simi Das; and my children, Sohan and Dikshan;

and my PhD supervisors, Dr. Sudeep Thepade and Prof. Saurav Ghosh.

I express my thankfulness to Xavier Institute of Social Service and all my colleagues

and collaborators for keeping faith in my abilities and for supporting

me throughout.

Contents

Preface

Mass popularity of expression of feelings and emotions with stunningly captured images, ideograms and smileys is no less than a pandemic in contemporary times. Applications such as Pixlr, Google Lens, etc. are default softwares in almost all our smart phones and PDAs. Currently, images are not only captured to create self portraits or vacation photographs, but also have far-reaching implications in terms of business value to organizations. Image data uploaded in public forums (social media) is used as a crucial source to understand dynamic consumer behavior online. It is also used to identify brands during online shopping and hence contributes directly to business revenue generation. However, in all these applications one thing is common: All the popular applications use image data by its content in terms of color, texture, shape etc. None of these require any text annotation to identify similar images online or offline. The reasons for developing these kinds of applications are simple. Users are interested in capturing images but not in tagging them professionally. Secondly, vocabulary may not be enough at times to describe image content.

This book is a comprehensive roadmap for step-by-step applications of content-based image identification techniques. The book thoroughly covers assorted techniques for implementing content-based image identification with enhanced precision. It illustrates the applications of diverse machine learning techniques to facilitate the process of content-based image recognition. The book is a guide to effective utilization of softwares and tools essential to pursue the process of content-based image recognition with high accuracy and less computational overhead.

Recent times have witnessed a surge in applying deep learning techniques to image data. Most content-based image classification tasks can be accurately executed using deep learning techniques, however these techniques are plagued by the following shortcomings:

- seldom perform well for small datasets
- require a huge amount of training data
- have high computational overhead
- require costly hardware
- are time-consuming

Apart from the aforesaid limitations, deep learning is more of a tool expertise than a domain expertise. For a learner, using deep learning techniques in the first go will never enable understanding of image processing basics necessary to fine-tune the performance of systems implementing content-based image classification. Moreover, the implementation overhead itself is a challenge for experimentation purpose with content-based image data.

This book readily addresses these shortcomings of the trending approaches and fully presents the basics of deep learning based approaches.

Primarily, the book covers various techniques of image processing including implementation of image binarization, image transforms, texture analysis and morphological approach. It explains the significance of all these techniques for robust feature extraction from image contents. Further, it demonstrates the process of content-based image identification with these features using popular machine learning algorithms. Finally, it also covers applications of deep learning techniques for feature extraction using representation learning. Hence, the book is an enabler toward domain expertise of the reader with a blend of application orientation with deep learning techniques.

I am sure the book will be a true companion to students, research scholars and industry professionals who wish to pursue a career in machine learning with content-based image data. It is helpful to spawn research ideas for PhD aspirants. It also provides the methodologies to design lightweight applications for identification using image contents.

Finally, I express my reverence to Almighty and love to my parents, wife and children for their continuous support and inspiration toward my work and experimentations. I express my gratitude toward my PhD supervisors, teachers, and collaborators from industry and academia, as well as my students and my colleagues for their encouragement and belief in my propositions.

Dr. Rik Das
Ranchi, India

MATLAB® is a registered trademark of The MathWorks, Inc. For product information, please contact:

The MathWorks, Inc.
3 Apple Hill Drive
Natick, MA 01760-2098 USA
Tel: 508-647-7000
Fax: 508-647-7001
E-mail: info@mathworks.com
Web: www.mathworks.com

Author

Dr. Rik Das

PhD (Tech.) in Information Technology , University of Calcutta

M.Tech. in Information Technology, University of Calcutta

B.E. in Information Technology, University of Burdwan

Assistant Professor, **PGPM (Information Technology), Xavier Institute of Social Service, Ranchi**

ACM Distinguished Speaker, Association for Computing Machinery, New York, USA, Professional Member, Association for Computing Machinery, New York, USA, Member of International Advisory Board, AI Forum, UK, Advisor, GradsKey (StartUp), SPOC, **SWAYAM—Local Chapter, XISS, Ranchi**

Rik is an assistant professor for the Programme of Information Technology at Xavier Institute of Social Service (XISS), Ranchi, India. He has more than 16 years of experience in academia and collaborative research with various leading universities and organizations in India and abroad.

Rik is appointed as a Distinguished Speaker of the Association of Computing Machinery (ACM), New York, USA, on topics related to Artificial Intelligence, Machine Learning, Computer Vision, and Natural language processing. He is featured in uLektz Wall of Fame as one of the "Top 50 Tech Savvy Academicians in Higher Education across India" for 2019. He is also a member of International Advisory Committee of AI-Forum, UK.

Rik is an invited speaker in multiple technical events, conclaves, meetups and refresher courses on Information Technology (machine learning, deep learning, image processing and e-learning) organized by prominent organizations such as the University Grants Commission (Human Resource Development Centre), The Confederation of Indian Industry (CII), Software Consulting Organizations, MHRD initiative under Pandit Madan Mohan Malviya National Mission on Teachers and Teaching, IEEE Student Chapters, Computer Science/Information Technology Departments of leading universities and many more.

Rik has collaborated with professionals from leading multinational software companies such as TCS, CTS etc. for research work in the domain of content-based image classification. His keen interest in the application of machine learning and deep learning techniques for designing computer-aided diagnosis systems for medical images has resulted in joint publication of research articles with professors and researchers from various universities abroad, including College of Medicine, University of Saskatchewan, Canada; Faculty of Electrical Engineering and Computer Science, VSB Technical

University of Ostrava, Ostrava, Czechia; Cairo University, Giza, Egypt and so on.

Rik has filed and published two Indian patents consecutively during 2018 and 2019 and has more than 40 international publications to date with reputed publishers. He has authored two books in the domain of content-based image classification and has edited three volumes on machine learning and deep learning with leading publishers. Many of his edited and authored volumes are currently in press awaiting release. Rik is an Editorial Board Member and Reviewer of multiple international journals of repute and has served as Organizing and Technical Committee member of several national and international conferences. He has also chaired sessions in international conferences on machine learning.

Rik started his career in 2002 as a business development executive at Great Eastern Impex Pvt. Ltd. (a bar code solution offering company). Afterwards, he joined Zenith Computers Ltd. as a senior marketing executive. His drive for higher education urged him to complete post-graduate and doctoral programs while pursuing a career in academics simultaneously. Rik has served Bengal Institute of Technology, Globsyn Business School and NMIMS (Mumbai) as a professor in the domains of computer science/information technology/information systems.

Rik earned his PhD (Tech) and MTech in information technology from the University of Calcutta, India. He also earned a BE in information technology from the University of Burdwan, India.

Rik founded a YouTube channel titled "Curious Neuron" to disseminate knowledge and information to larger communities in the domain of machine learning, research and development and open source programming languages. He is always open to discuss new research project ideas for collaborative work and for techno-managerial consultancies.

1

Introduction to Content-Based Image Classification

1.1 Prelude

A picture collage contains an entire life span in a single frame. We have witnessed global excitement with pictorial expression exchanges compared to textual interaction. Multiple manifestation of social networking have innumerable image uploads every moment for information exchanges, status updates, business purposes and much more [1]. The cell phone industry was revolutionized with the advent of camera phones. These gadgets are capturing high-end photographs in no time and sharing the same for commercial and noncommercial usage [2]. Significant medical advancements have been achieved by Computer-Aided Diagnosis (CAD) of medical images [3]. Therefore, image data has become inevitable in all courses of modern civilization, including media, entertainment, tourism, sports, military services, geographical information systems, medical imaging and so on.

Contemporary advancement of computer vision has come a long way since its inception in 1960 [4,5]. Preliminary attempts were made for office automation tasks pertaining to approaches for pattern recognition systems with character matching. Research work by Roberts has envisaged the prerequisite of harmonizing two-dimensional features extracted from images to three-dimensional object representations [6]. Escalating complexities related to unevenly illuminated pictures, sensor noise, time, cost, etc. have raised realistic concerns for continuing the ensuing research work in the said domain with steadfastness and uniformity.

Radical advancements in imaging technology have flooded the masses with pictures and videos of every possible detail in their daily lives. Thus, the creation of gigantic image datasets becomes inevitable to store and archive all these rich information sources in the form of images. Researchers are facing mounting real-time challenges to store, archive, maintain, extract and access information out of this data [7].

Content-based image classification is identified as a noteworthy technique to handle these adversities. It has been considered effective to identify image data based on its content instead of superficial annotation.

Image annotation is carried out by labeling the image content with text keywords. It requires considerable human intervention to manually perform this action. Moreover, the probability of erroneous annotation is high in cases of labeling gigantic image datasets with a manual text entry procedure. The text tag describing image content is as good as the vocabulary of the person who tags it. Thus, the same image can have different descriptions based on the vocabulary of the annotation agent responsible for it, which in turn hampers the consistency of the entire process [8].

Conversely, extraction of a feature vector from the intrinsic pixels of the image data has eradicated the challenges faced due to manual annotation and has automated the process of identification with minimal human intervention [9]. Present day civilization follows the trend of capturing images of events and objects of unknown genre. The process of content-based image identification can readily classify the captured images into known categories with the help of preexisting training knowledge. This, in turn, assists in decision-making for further processing of the image data in terms of assorted commercial usage.

Promptness is not the only decisive factor for efficient image classification based on content. Accuracy of classification results contribute immensely to the success factor of a classification infrastructure. Thus, to ensure the competence of content-based image classification, one has to identify an effectual feature extraction technique. The extracted features become pivotal to govern the success rate of categorizing the image data into corresponding labels.

Therefore, different feature extraction techniques are discussed in this work to represent the image globally and locally by means of extracted features. The local approach functions on segmented image portions for feature extraction, contrary to the global approach. However, image data comprises a rich feature set, which is seldom addressed by a single-feature extraction technique. As a result, fusion of features has been explored to evaluate the classification results for improved accuracy.

The experiments are carried out on four widely used public datasets using four different classifiers to assess the robustness of extracted features. Diverse metrics, such as Precision, Recall, Misclassification Rate (MR) and F1 Score, are used to compare the classification results. A brief explanation of each of the metrics used is given in the following section. It is followed by the description of the classifiers and the datasets.

1.2 Metrics

Different parameters have been used to measure the classification performances of the diverse feature extraction techniques [10]. A brief explanation for each of the parameters is also provided.

1.2.1 Precision

Precision is defined as the fraction of appropriate classification among the classified instances as in equation 1.1

$$Precision = \frac{TP}{TP + FP} \tag{1.1}$$

1.2.2 True Positive (TP) Rate/Recall

Recall is defined as the fraction of appropriate classification among the total number of related instances as in equation 1.2.

$$TP\ Rate/Recall = \frac{TP}{TP + FN} \tag{1.2}$$

1.2.3 Misclassification Rate (MR)

Misclassification Rate (MR) is defined as fraction of incorrectly classified instances and is denoted as the error rate of the classifier as in equation 1.3.

$$MR = \frac{FP + FN}{TP + TN + FP + FN} \tag{1.3}$$

1.2.4 F1-Score

The weighted average of Precision and Recall (TP Rate) is deduced as the F1 Score of classification and is given as in equation 1.4.

$$F1\ Score = \frac{2 * Precision * Recall}{Precision + Recall} \tag{1.4}$$

1.2.5 Accuracy

Accuracy of a classifier is measured by means of the recognition rate of the classifier for identifying the instances correctly and is given as in equation 1.5.

$$Accuracy = \frac{TP + TN}{TP + TN + FP + FN} \tag{1.5}$$

1.2.6 False Positive (FP) Rate

This is the fraction of incorrect results classified as positive results by the classifier and is given as in equation 1.6.

$$FP\ Rate = \frac{FP}{FP + TN} \tag{1.6}$$

1.2.7 True Negative (TN) Rate

This metric is the fraction of negative results correctly produced for negative instances and is given as in equation 1.7.

$$TN = \frac{TN}{FP + FN} \tag{1.7}$$

1.2.8 False Negative (FN) Rate

This metric provides the fraction of positive results that are classified as negative results by the classifier and is given as in equation 1.8.

$$FN\ Rate = \frac{FN}{TP + FN} \tag{1.8}$$

1.3 Classifiers

Diverse feature extraction techniques have been implemented to assess the classification performance under varied classifier environments. This is done to examine the consistency of the corresponding feature extraction technique when applied to different datasets. Four different types of classifiers have been considered to carry out the task. The classifiers are K Nearest Neighbor (KNN) Classifier, Ripple-Down Rule (RIDOR) Classifier, Artificial Neural Network (ANN) Classifier and Support Vector Machine (SVM) Classifier. Brief descriptions of the classifiers explain their individual work flows [10].

1.3.1 KNN Classifier

The highest error rate for K Nearest Neighbor (KNN) is twice the Bayes error rate, which is considered as the minimum possible rate for an instance-based classifier. Thus, it is considered as one of the classifiers for evaluation purposes. It classifies the unknown instance by identifying its nearest neighbor in the instance space and further designates the class of the nearest neighbor to the unknown instance.

The working formula of four different similarity measures, namely, Euclidean Distance, Canberra Distance, City Block Distance and Mean Squared Error, have been given in equations 1.9–1.12. These similarity measures are implemented to figure out the nearest neighbors during the classification process of the instances.

$$D_{Euclidian} = \sqrt{\sum_{i=1}^{n} (Q_i - D_i)^2} \qquad (1.9)$$

$$D_{Canberra} = \sum_{i=1}^{n} \frac{|Q_i - D_i|}{|Q_i| + |D_i|} \qquad (1.10)$$

$$D_{City\ block} = \sum_{i-1}^{n} |Q_i - D_i| \qquad (1.11)$$

$$D_{MSE} = \frac{1}{n} \sum_{i=1}^{n} (Q_i - D_i)^2 \qquad (1.12)$$

1.3.2 Random Forest Classifier

It is a classifier which acts on the principle of ensemble learning. It builds a number of decision trees during the time of training and the output class is determine by the mode of the classes it learns during the supervised procedure.

Random Forest Classifier has a number of advantages that have made it a popular choice for the classification task. It can proficiently manage large datasets by handling thousands of input variables. It refrains from variable deletions and approximates the significance of a variable for classification.

It efficiently estimates missing data to maintain classification accuracy. The forest generated during the classification process can be saved for future use on different datasets.

1.3.3 ANN Classifier

Complex pattern recognition is carried out with the Artificial Neural Network (ANN) Classifier, which is a mixture of simple processing units. Enhanced noise tolerance level and properties of self-adaptation have made ANN classifier proficient for real-time applications. A category of feed forward artificial neural network named multilayer perceptron (MLP) is used in this work, which has implemented back propagation for supervised classification. Classification performance is optimized by training the network using feed forward technique. A predefined error function is computed by comparing the predicted values with the correct answers. Weight adjustment of each connection is done henceforth by feeding the errors to the network to reduce the error values. The overall error is reduced in this way by executing the process repeatedly for a large number of training cycles. The predefined error function is calculated by using the back propagation function with known or desired output for each input value. An MLP framework is comprised of input nodes, output nodes and hidden nodes. The number of input nodes is determined based on the summation of the number of attributes in the feature vector and the bias node. Output nodes are equal to the number of class labels. The hidden nodes are structured based on a predefined value. Multiple layers in the classification framework have the output flow toward the output layer, as in Fig. 1.1. The MLP shown in Fig. 1.1 has two inputs and a bias input with weights 3, 2, and −6 respectively. The activation function f_4 applied to value S is given by $S = 3x_1 + 2x_2 - 6$.

Unipolar step function is used to calculate the value of f_4 given in equation 1.13. It has used an output of 1 to classify into one class and an output of 0 to pass in the other class.

$$f_4 = \begin{cases} 1 & if \ldots S > 0 \\ 0 & otherwise \end{cases} \tag{1.13}$$

An alternative representation of the values of 3, 2, and −6 as respective weights of three different inputs for the perceptron is shown in Fig. 1.2.

The horizontal and vertical axis is denoted by x_1 and x_2 respectively. The intercepting coordinates of the straight line to the vertical and the horizontal axes are the weights. Two classes are categorized by the area of the plane on the left and the right side of the line $x_2 = 3 - 3/2x_1$.

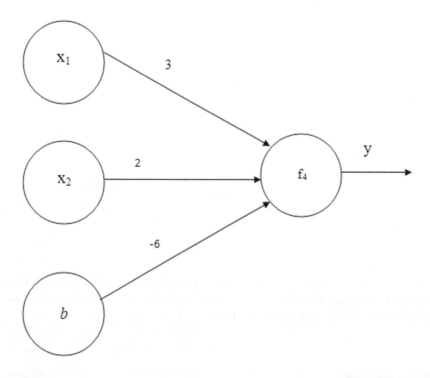

FIGURE 1.1
Multilayer Perceptron.

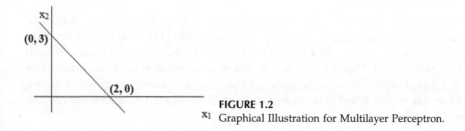

FIGURE 1.2
Graphical Illustration for Multilayer Perceptron.

1.3.4 SVM Classifier

Nonlinear mapping is used by Support Vector Machine (SVM) Classifier to convert original training data to a higher dimension. Searching of Optimal separating hyperplane is carried out in this new dimension. The hyperplane is useful to separate data from two different classes using an appropriate nonlinear mapping to an adequately high dimension. The methodology is illustrated in Fig. 1.3, where support vectors are denoted with thicker borders.

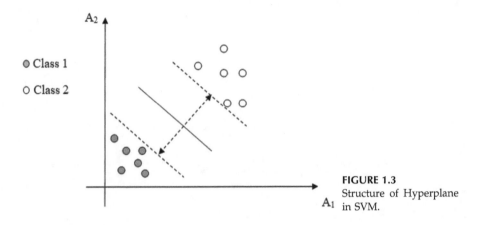

FIGURE 1.3
Structure of Hyperplane in SVM.

The idea was to implement the feature extraction techniques with widely used public datasets namely, Wang Dataset, Caltech Dataset, Corel Dataset and Oliva-Torralba (OT-Scene) Dataset.

1.4 Datasets Used

1.4.1 Wang Dataset

This popular public dataset is offered by Wang et al. It is divided into 10 categories comprising 1000 images on the whole [11]. The dimensionality of each image is 256×384 or 384×256, and the number of images in each category is 100. The dataset is divided into categories such as Tribal, Sea Beaches, Gothic Structures, Buses, Dinosaur, Elephants, Flowers, Horses, Mountains and Food. The Wang dataset is illustrated in a sample collage in Fig. 1.4.

1.4.2 Caltech Dataset

Caltech dataset has 8,127 images spread across 100 categories [12]. The categories include an unequal number of images individually. The dimension of each image is 300×200. Different categories include accordion, airplanes, anchor, ant, background Google, barrel, bass, beaver, binocular, bonsai, brain, brontosaurus, Buddha, butterfly, camera, cannon, car side, ceiling fan, cellphone, chair, etc. The Caltech dataset is illustrated in a sample collage in Fig. 1.5.

FIGURE 1.4
Sample Wang Dataset.

1.4.3 Corel Dataset

Corel dataset embraces 10,800 images of dimension 80×120 or 120×80 spread across 80 different categories [13]. The categories are named art, antique, cyber, dinosaur, mural, castle, lights, modern, culture, drinks, feast, fitness, dolls, aviation, balloons, bob, bonsai, bus, car, cards, decoys, dish, door, Easter eggs, faces etc. The Corel dataset is illustrated in a sample collage in Fig. 1.6.

1.4.4 Oliva Torralba (OT-Scene) Dataset

OT-Scene is a popular public dataset offered by MIT [14] having 8 categories of images totalling 2688 images. Assorted categories in the dataset are named Coast and Beach (360 images), Open Country (328 images), Forest (260 images), Mountain (308 images), Highway (324 images), Street (410 images), City Centre (292 images) and Tall Building (306 images). A collage illustration of OT-Scene dataset is shown in Fig. 1.7.

FIGURE 1.5
Sample Caltech Dataset.

FIGURE 1.6
Sample Corel Dataset.

FIGURE 1.7
Sample OT-Scene Dataset.

1.5 Organization of the Book

Currently, this work attempts to furnish a foretaste of possibilities of applying multiple innovative techniques to stimulate the performance of content-based image classification. Each of the methods is discussed in detail in the forthcoming chapters, with special emphasis on creating a framework for fusion-based approaches.

Chapter 1 provides an overview of current scenarios for content-based image classification. The chapter emphasizes the importance of feature extraction techniques to ensure the success of content-based image recognition.

Chapter 2 discusses various contemporary works on benchmarked feature extraction techniques. It reviews the recent techniques reported in content-based image classification and discusses the importance of various techniques for enhanced classification results.

Chapters 3–7 discuss various handcrafted techniques of feature extraction for content-based image classification. The chapters specifically describe various methodologies to extract features from low-level image characteristics such as color, shape, texture, etc. The author provides a step-by-step guide for hands-on experience for the reader in implementing the algorithms. The extracted features are significantly small in dimension, which help the reader test the robustness of the extracted features—even on a system with basic configurations.

Chapter 8 establishes the significance of classification using a fusion frame-work for increased classification results. The chapter makes an efficiency comparison of early fusion to late fusion and establishes the importance of fusion-based architecture for enhanced classification results.

Finally, Chapter 9 explains representation learning with pretrained Convolutional Neural Network (CNN) architecture. The book concludes with direction toward future scope and diverse applications possible with the help of content-based image classification.

Chapter Summary

The chapter has introduced the concept of content-based image classification. It has discussed the background and objectives of classification based on image content. The metrics used throughout this book for evaluating the efficacy of the classification process are also illustrated in this chapter. Finally, the chapter has demonstrated the different datasets used in this book for explaining the various schemes of feature extraction for content-based image classification.

References

1. Hu, J., Yamasaki, T. and Aizawa, K., 2016, May. Multimodal learning for image popularity prediction on social media. In 2016 IEEE International Conference on Consumer Electronics-Taiwan, ICCE-TW, IEEE, pp. 1–2.
2. DeBerry-Spence, B., Ekpo, A. E. and Hogan, D., 2019. Mobile phone visual ethnography (MpVE): Bridging transformative photography and mobile phone ethnography. *Journal of Public Policy & Marketing*, 38(1): 81–95.
3. Ma, J., Wu, F., Jiang, T. A., Zhao, Q. and Kong, D., 2017. Ultrasound image-based thyroid nodule automatic segmentation using convolutional neural networks. *International Journal of Computer-Assisted Radiology and Surgery*, 12(11): 1895–1910.
4. Roberts, L. G., 1960. Pattern recognition with an adaptive network. In *Proceedings of IRE International Convention Record*, pp. 66–70.
5. Tippett, J. T., Borkowitz, D. A., Clapp, L. C., Koester, C. J. and Vanderburgh, A. J. (Eds.), 1965. *Optical and Electro-Optical Information Processing*, MIT Press.
6. Roberts, L. G., 1963, Machine Perception of Three Dimensional Solids, Ph.D. thesis, Massachusetts Institute of Technology.

7. Andreopoulos, A. and Tsotsos, J. K., 2013. 50 Years of object recognition: Directions forward. *Computer Vision and Image Understanding*, 117(8): 827–891.
8. Sohn, S. Y., 1999. Meta analysis of classification algorithms for pattern recognition. IEEE Transactions on Pattern Analysis and Machine Intelligence, 21(11): 1137–1144.
9. Kekre, H. B., Thepade, S., Kakaiya, M., Mukherjee, P., Singh, S. and Wadhwa, S., 2011. Image retrieval using shape texture content as row mean of transformed columns of morphological edge images. *International Journal of Computer Science and Information Technologies*, 2(2): 641–645.
10. Sridhar, S., 2001. *Digital Image Processing*, Oxford University publication.
11. http://wang.ist.psu.edu/docs/related/ (accessed March 18, 2020).
12. http://www.vision.caltech.edu/Image_Datasets/Caltech101/ (accessed March 18, 2020).
13. https://sites.google.com/site/dctresearch/Home/content-based-image-retrieval (accessed March 18, 2020).
14. https://lmb.informatik.uni-freiburg.de/resources/datasets/SceneFlowDatasets. en.html#downloads (accessed March 18, 2020).

2

A Review of Handcrafted Feature Extraction Techniques for Content-Based Image Classification

---

2.1 Prelude

An enormous quantity of valuable information is accessible from the image data archived in colossal databases. Assorted sources, including entertainment media, mainstream media, defense, sports events and social media platforms, contribute readily to form these image data archives [1]. Numerous approaches are suggested and carried out in recent years with the purpose of extracting valuable information from these enormous data and classifying them into corresponding categories. This chapter assesses and discusses some of the significant allied approaches in the aforementioned domain of feature extraction and classification techniques for content-based image classification.

Exploring information present in an image is traditionally carried out by implementing diverse feature extraction techniques on low-level image characteristics such as, color, texture and shape [2]. The two most important stages in content-based image classification consist of robust feature extraction from image data followed by classification of image categories with the help of extracted signatures.

A concise study of significant methods for extracting content-based image descriptors (features) and of benchmarked classification techniques for analyzing the performance of the extracted descriptors is described in the following subsections.

2.2 Extraction of Features with Color Contents

One of the prominent characteristics of an image is considered to be the color components for extraction of meaningful features for content-based image classification (CBIC). One of the approaches has implemented a technique which considers three horizontal regions in the image that do not overlap, and from each of the zones, the first three moments for each separate color channel is extracted to a feature vector of 27 floating points [3]. Some of the approaches are observed to consider three major attributes for extraction of color-based characteristics, namely, color space, quantization of the color space and extraction of the color feature [4–6]. Color features have explored different methods like conventional color histogram (CCH) and fuzzy color histogram (FCH) [7,8].

CCH has represented frequency of occurrence of each color component in an image. It has the benefit of uncomplicated implementation that results in reduced computational overhead. One major drawback of the technique is its bulky feature dimension, which stays identical irrespective of thorough quantization of color space. Another deficiency also detected is its incapability to assess similar color across different bins. Spatial information of color spatial is also not encoded in CCH.

However, different extent of memberships for each bin is observed in case of FCH because it belongs to all histogram bins. A primary advantage of FCH is its capability to encode the level of similarity of each pixel color to all other histogram bins by implementing a fuzzy-set membership function. FCH is proven to be additionally robust in handling quantization errors for varying light intensity. Nevertheless, the curse of dimensionality has turned out to be the primary reason for its setback supplemented with increased computational complexity of fuzzy membership function (see Fig. 2.1).

2.3 Extraction of Features with Image Binarization

Segregation of the required content in the picture from its background for proficient extraction of its characteristics is rightly carried out with binarization method. The method of image binarization has comprehensively implemented selection of appropriate threshold to differentiate the region of interest in an image from its background. In recent years, numerous techniques are proposed by the researchers for extraction of binarized features using threshold selection [9–11]. Binarization technique is also found to be useful to carry out image descriptor definition as discussed in existing literature [12–19]. However, most of the proposed methodologies are observed to be evaluated in controlled environments facilitated by processing professional-quality images available from high-end image-capturing devices. On the contrary, capturing images in a real-life scenario may be affected by diverse factors comprising of low-light condition, nonuniform illumination, etc. [20]. Proper threshold selection technique becomes imperative in such a scenario to identify the foreground object separately from

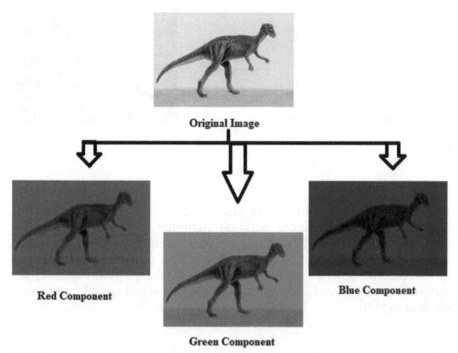

FIGURE 2.1
RGB Model of Color Components.

the background in those pictures. Crucial challenges for threshold selection in binarization technique are governed by altering illumination and noise [21]. Variance of gray levels present in the object and the background, inadequate contrast, etc. can adversely influence the calculation of threshold for binarization [22]. Choice of threshold is of utmost importance because a flawed threshold value may misinterpret a pixel from the background as foreground and vice versa. The process of threshold selection is primarily divided into three different categories, namely, mean threshold selection, global threshold selection and local threshold selection. Variety of mean threshold selection techniques are efficiently formulated for extraction of important image descriptors for CBIC [23,24]. But, the technique of mean threshold selection has not accounted for the standard deviation of the distribution of the gray level values in an image. Image variance as a component of image threshold selection is explored with traditional Otsu's method of global threshold selection [25,26]. Techniques based on entropy for threshold selection are suggested for extraction of robust descriptors [27]. Adaptive threshold selection for binary image creation has also considered contrast of an image as a parameter in addition to variance and mean of the pixel values (Fig. 2.2). A prominent example of successful execution of adaptive threshold selection is Niblack's algorithm for adaptive

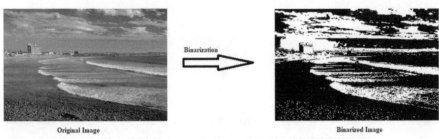

Original Image   Binarized Image

FIGURE 2.2
Process of Image Binarization.

threshold selection for images of veins present in a human finger [28]. Sauvola's adaptive threshold selection algorithm is considered to be an improvement over Niblack's technique due to its efficiency in the domain of optical character recognition [29]. A nonuniform illumination problem of quick response (QR) image is dealt with in Bernsen's local threshold selection method [30,31]. The optimal threshold is calculated with error method in [32].

2.4 Extraction of Features with Image Transforms

Image transformation leverages shifting of the domain of the image elements, which results in representation of an image as a set of the energy spectrum. A basis image is created by extrapolation of the image to a series of basis functions [33]. The basis images are generated by means of orthogonal unitary matrices in the form of image transformation operators. Transformation of an image from its current representation to a different one has two different benefits:

a. Expansion of an image is possible as a waveform series by using image transforms. The process of transformation is also helpful in distinguishing the critical components of image patterns.

b. Information can be stored in a compact manner within the transformed image data, which increases the efficiency and usefulness of the process.

The aforementioned characteristics of image transform support a thorough reduction of descriptor dimension extracted from the images using diverse transformation techniques, namely, discrete cosine transforms, Hartley transform, etc. Therefore, the researchers have proposed different techniques of feature extraction by utilizing the properties of image transforms to extract features from images using fractional energy coefficient [34,35], row mean of column transformed image [36,37], energy compaction [38] and Principle Component Analysis (PCA) [39]. The techniques have considered seven image transforms and fifteen fractional coefficient sets for efficient feature extraction. Original images are divided into subbands by using multiple scales Biorthogonal wavelet transform, and the subband coefficients are used as features for image classification [40]. The feature spaces are reduced by applying Isomap-Hysime random anisotropic transform for classification of high dimensional data [41]. Important texture features are identified as features by using wavelet transform for Glaucomatous image classification [42]. Spatial features are extracted using stationary wavelet transform for hyperspectral image feature classification [43]. The property of multilevel decomposition of wavelet transform is used for feature extraction from nailfold capillary images [44]. Dirichlet Fisher kernel is introduced for transforming the histogram feature vector to enhance the classification results [45]. Microscopic images are well classified by shearlet transform [46]. Shearlet has intrinsic directional sensitivity in contrast to wavelet filters such as the Gabor filter. This made it appropriate for differentiating small contours of carcinoma cells. Visual information is captured by shearlet transform by applying a multiscale decomposition, provided by edges detected at different orientations and multiple scales. The Fukunaga-Koontz transform (FKT) is implemented as an efficient target detection method useful for remote sensing tasks, and it is principally used as an effective classification tool for two-class problems. Increased nonlinear discrimination ability and capturing higher order of statistics of data has been made possible by kernelizing the traditional FKT [47]. Discrete cosine transform has been useful in extracting features for face recognition in which a high success rate is achieved with a lesser amount of data [48]. A nonparametric local transform, named census transform, is used for establishing relationships among local patches to extract feature vectors for image recognition [49]. Wavelet packets are used to generate feature vectors for content-based image identification.

2.5 Extraction of Features with Morphological Processing

The contour of the objects present in an image can be resourcefully analyzed by mathematical morphology, which is based on set theory. A binary image is considered as a set, and application of set operators, namely, intersection, union, inclusion and complement, can be well performed on the image [50]. Representation of shape can be done based on boundary or region. Boundary-based shape representation corresponds to the outer boundary, which is extracted as edges by gradient and morphological operators. Edges in only one direction can be determined (horizontal, vertical or diagonal) by application of gradient operator, which only provides the first order derivative [33]. Slope magnitude method is used along with gradient operators [51] to define the complete outside edge of the shape in the image by creating an outline of connected edges. Local descriptors [52–54] of shape are prioritized over the global descriptors [55] with passing time due to distinctive modeling restrictions. Representation of shape by discrete curve evolution is helpful to simplify contours to eliminate noisy and irrelevant shape attributes [56]. A new shape descriptor, known as shape context, is recognized by multiple geometric transformations as a convincingly compact and robust means for feature extraction [57]. A metric tree is arranged for feature representations (curvature and orientation) [58] by expressing the curves with a set of segments or tokens [53]. The purpose of shape matching is fulfilled by approximating the shapes as sequences of concave and convex segments [54]. Investigation of morphological profiles (MP) is conducted with partial reconstruction and directional MPs for the categorization of high-resolution hyperspectral images from urban areas [59]. Medical X-ray image classification is facilitated by extraction of shape features such as Fourier Descriptor, Invariant Moments, and Zernike Moments from the image [60]. Regions of interest (ROIs) from the input image of the traffic signals for traffic sign classification are distinguished by the use of morphological operators, segmentation and contour detection [61]. Automated damaged building detection is carried out by applying morphological operations of opening and closing for segmented images [62]. Image enhancement has also used morphological operator prior to image segmentation of pavement images to detect pavement cracking [63]. Leaf image classification has used shape context as a global feature [64]. Accurate classification is observed with images based region features calculated by eigenregions. Shape features capable of describing the 3-D shape of the target objects are incorporated for supervoxel-based segmentation of mitochondria [65]. An elaborate study on Fourier descriptor, curvature scale space (CSS), angular radial transform (ART) and image moment descriptors for shape representation is carried out for shape-based image recognition [66]. Shape feature is efficiently represented by designing low-level shape descriptors, which has added robustness to rotation, scaling and deformation of shapes [67]. Midlevel modeling of shape representation is focused by developing a new shape demonstration called Bag of Contour Fragments (BCF) in which a shape is decomposed into contour fragments and is described individually using a shape descriptor (Fig. 2.3).

Classification of contour shapes using class contour segments as input features with Bayesian classifier is proposed which has designed a shape classification framework [68]. Input feature comprising of contour segments and skeleton paths are considered for categorization of shape with a Gaussian mixture model [69]. Transformation of contour points into a

Original Image   **Morphological Operation**

FIGURE 2.3
Applying Morphological Operator.

symbolic representation followed by the editing of distance between a pair
of strings is used to classify with a kernel support vector machine [70].
Prototyping with a tree-union representation for each shape category for
shape classification is performed [71]. A skeletal tree model has represented
the category-wise prototype for image classification [72]. A shape codebook
is represented by each prototype [73]. Skeleton matching is useful to study
the shape classification [74–77]. Contour-based methods have included CSS
[78], triangle area representation (TAR) [79], hierarchical procrustes
matching (HPM) [80], shape-tree [81], contour flexibility [82], shape context
(SC) [83], inner-distance shape context (IDSC) [84] and so on.

2.6 Extraction of Features with Texture Content

Texture in images is formed due to variations in the intensity and color. It is
also present in the form of repeated patterns. Uniformity or placement rules
form the basic primitives of repetitive patterns. Textures correspond to qualitative terms such as coarseness, homogeneity, and smoothness in the physical
level. Based on the size of the primitives, textures are of two types, micro
textures and macro textures [85]. Macro texture or coarseness is formed by the
larger texture of the object, and the smaller textures lead to fine or micro textures. The orientation of the texture elements is called directionality.
Classification can be performed based on the nature of the surface as smooth or
coarse. The texture features are used to propose adaptive local binary patterns
for image classification [86]. Segmentation based on wavelet statistical texture
features is used for classification of computed tomography images of the brain
[87]. A novel texture pattern representation method called Hierarchical Visual
Codebook (HVC) is proposed to encode the texture primitives of iris images for
iris image classification [88]. Important information for classification can be
extracted from texture-based analysis of land covers exhibiting customary
patterns in very high resolution remotely sensed images [89].

Various methods of texture-based feature extraction techniques comprised the steerable pyramid, the contourlet transform, the Gabor wavelet transform [90] and the complex directional filter bank (CDFB) [91]. The shift-invariant CDFB is introduced for texture based image identification [91]. The significance of CDFB consists of shift invariance, relatively enhanced texture retrieval performance and comparatively minimized redundancy ratio. The over-complete ratio of the CDFB has been restricted to 8/3, compared to those of the Gabor transform and steerable pyramid that increases linearly with the number of directional sub-bands [91]. A multiscale, multidirectional representation of the image can be created by steerable pyramid. The translations and rotations of a single function formulate the basic filters. Representation of texture can be categorized in three different ways, namely structural, statistical and multiresolution filtering methods. Utility of texture features have been investigated widely in the past for its applications in image processing, computer vision and computer graphics [92] such as multiorientation filter banks [93] and wavelet transforms [94–96]. Statistics of coefficients of certain transforms obtained from application on the image pixels can be used to generate texture features. Generation of texture features by application of wavelet transform and the discrete cosine transform has been achieved in two different occassions mentioned in literature [97] [98]. Computer vision and graphics have made commendable progress in the areas of texture synthesis, in which Markov statistical descriptors based on pairs of wavelet coefficients at adjacent location/orientation/scale in the images are used [99]. Manjunath and Ma [100,101] contributed significantly at the initial stage of working with texture features for image recognition and indicated the appropriateness for inclusion in the MPEG-7 [102]. Ma and Manjunath have also proposed [103] a thesaurus for texture, meant for aerial image identification. Extraction of texture involved in the thesaurus-building process is based on the application of a bank of Gabor filters [104] to the images to encode statistics of the filtered outputs as texture features. Textured region descriptors such as affine- and photometric-transformation invariant features are observed to be robust to the shape of the region in question [105]. Affine invariant texture feature extraction conceived for advancement in texture recognition is observed in contemporary literature [106] (Fig. 2.4).

2.7 Fusion of Features Extracted with Multiple Techniques

Color moments and moments on Gabor filter responses are considered for extraction of local descriptors from color and texture. Shape information is taken out in terms of edge images using gradient vector flow fields. The shape features are ultimately represented by invariant moments. The content-based image recognition decision and all these individual features are fused for enhanced retrieval performance [107]. Color histogram-based feature vectors and

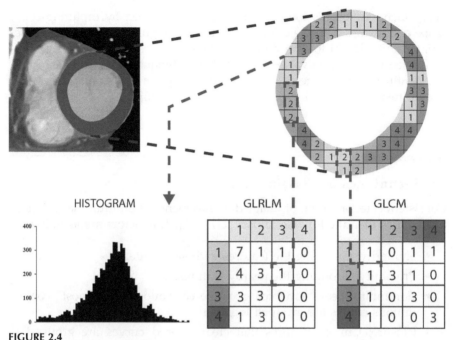

FIGURE 2.4
Texture Analysis of Heart Muscle.
Source: Hinzpeter, R., Wagner, M.W., Wurnig, M.C., Seifert, B., Manka, R. and Alkadhi, H., 2017. Texture analysis of acute myocardial infarction with CT: First experience study. PloS One, 12(11).

a co-occurrence matrix-based texture features are extracted from hue, saturation, value (HSV) color space for content-based image identification [108]. Fuzzy set theoretic approach is implemented to choose visually significant point features from images. These points are used to compute some invariant color features to calculate the similarity between images [109]. Fusion of color layout descriptor and Gabor texture descriptor as image signatures has portrayed increased image identification results [110]. Multitechnique feature extraction comprising color, texture and spatial structure descriptors has resulted in augmented recognition rate [49]. Neural network architecture of image identification is formed to evaluate content-based similarity matching of images using wavelet packets and Eigen values of Gabor filters as feature vectors [111]. Intraclass and interclass feature extraction from images have incorporated improved recognition results with image data [112]. Color co-occurrence matrix (CCM) and difference between pixels of scan pattern (DBPSP) have been exploited for extraction of color and texture-based features that are given as input to an artificial neural network (ANN) based classifier [113]. In [114], integration of two techniques of feature extraction embracing modified color motif co-occurrence matrix (MCMCM) and DBPSP features with equal weights has proved to be resourceful for content-based image identification. Fusion of retrieval results obtained by capturing color, shape and texture with the color moment (CM), angular radial transform descriptor and edge histogram descriptor (EHD) features respectively has outperformed the Precision values of individual techniques [115]. Unique feature are identified from images by using Color histogram and spatial orientation tree for fusion framework of image identification [116].

2.8 Techniques of Classification

Consistent categorization of image data is dependent on multiple factors in common as identified by Gestaltists [117–121]. The factors are as follows:

- Similarity: Grouping the features with similitude.
- Proximity: Grouping the nearby features.
- Continuity: Features which generate continuous or "almost" continuous curves are grouped together.
- Closure: Curves/features that create closed curves are grouped together.
- Common fate: Grouping of features with coherent motion.
- Parallelism: Grouping of parallel curves/features.
- Familiar configuration: Features whose grouping leads to familiar objects are usually grouped together.
- Symmetry: Grouping of curves that create symmetric groups.
- Common region: Grouping of feature lying inside the same closed region.

Assorted algorithms are proposed for the process of image classification. The following subsections briefly discuss each of the techniques of learning methods for classifications.

2.9 Logic-Based Algorithms

2.9.1 Decision Trees

Instances are classified by the decision trees by sorting them based on feature values. A feature in an instance to be classified is signified by each node in a decision tree, and the assumed value of the node is represented by the corresponding branch. Classification of instances is initiated at the root node, and the sorting has to be done with respect to the feature values. The root node of the tree comprises the feature that best divides the training data. Multiple techniques are applied to recognize the feature that best divides the training data such as information gain and Gini index [122,123]. However, each attribute is measured independently by the myopic measures; namely, ReliefF algorithm [124] estimates them in the context of other attributes. Nevertheless, it is observed by majority of studies that there is no single best method [125]. It is still vital to compare individual methods for deciding which metric should be used in a particular dataset. The same procedure is then repeated on each partition of the divided data, creating subtrees until the training data is divided into subsets of the same class. Splits based on a single feature at each internal node are used by decision trees, which characterize them as univariate. The problems involving diagonal partitioning cannot be handled well by most of the decision trees. The division of the instance space is orthogonal to the axis of one variable and parallel to all other axes. Hence, all hyper-rectangular regions are obtained after partitioning. However, some of the methods construct multivariate trees. Zheng's method has improved the accuracy of classification for the decision trees by creating new binary features with logical operators such as conjunction, negation and disjunction [126]. In addition, Zheng created at-least M-of-N features [127]. The value of a minimum M-of-N representation for a given instance is true if at least M of its conditions is true for the instance; otherwise, it is false. A combined decision tree is formed with a linear discriminant for constructing multivariate decision trees [128].

The most familiar algorithm in the literature for construction of decision trees is the C4.5 [129]. Comprehensibility is considered as one of the most important features of decision trees. The assumption made in the decision trees is that instances belonging to different classes have different values in at least one of their features. Decision trees tend to perform better when dealing with discrete/categorical features (Fig. 2.5).

2.9.2 Learning a Set of Rules

A variety of rule-based algorithms can be directly induced from the training data [130]. Each class is represented by disjunctive normal form (DNF) in classification rules. The objective is to create the smallest ruleset consistent with the training data. Learning a large number of rules signifies that the learning algorithm has been trying to "remember" the training set instead of discovering the assumptions that governed it. A separate-and-conquer algorithm (covering algorithms) search for a rule that explains a part of its training instances separates these instances and recursively conquers the remaining instances by learning more rules until no instances remain. The heuristics for rule learning and for

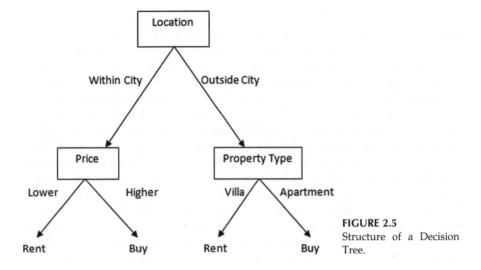

FIGURE 2.5
Structure of a Decision Tree.

decision trees are differentiated by their evaluation property. The latter one assesses average quality of a number of disjointed sets, and the former only gauges the quality of the set of instances that is covered by the candidate rule. It is important for a rule induction system to generate decision rules that have high predictability or reliability; this property is measured by a function called rule quality. Both the rule induction and classification processes require a rule-quality measure such as J-measure [131]. Rule induction has implemented the rule-quality measure as a decisive factor in the rule designing and/or simplification process. Classification has combined a rule-quality value with each rule to resolve conflicts by satisfying multiple rules by the example to be classified. A number of statistical and empirical rule-quality measures are surveyed [132]. An analysis of the performance of separate-and-conquer or covering rule-learning algorithms is performed by visualizing their evaluation metrics [133]. RIPPER is a well-known rule-based algorithm [134]. Repeated growing and pruning is followed by this algorithm to grow rules. The growing phase is involved in preparing restrictive rules in order to best fit the training data. The pruning phase is meant to make the rules less restrictive to avoid overfitting, which reduces performance on unseen instances. Hence, the distinguishing feature of rule-based classifiers is their unambiguousness. Moreover, the experts have proposed discritization of features before induction to increase classification accuracy with reduced training time [135]. Improved classification accuracy of rule learning algorithms is observed by combining features (such as in decision trees) by using the background knowledge of the user [136] or automatic feature construction algorithms [137].

2.9.3 Perceptron-Based Techniques

2.9.3.1 Single-Layer Perceptrons

A single-layer perceptron is briefly described as follows: If the input feature values be denoted by $x1$ through xn and the connection weights/prediction vector be denoted by w_1 through w_n (typically real numbers in the interval [−1, 1]), then the sum of weighted inputs is calculated by the perceptron as: and $\sum_i x_i w_i$ an adjustable threshold is maintained for the output. Output is assigned with a value of 1 if the sum is above the threshold; else it is 0. A frequently used learning process for the perceptron algorithm from a batch of training instances is to repeat the execution of the algorithm through the training set to search for an appropriate prediction vector for the entire training set. Label prediction of the test set is executed by this prediction rule. A novel algorithm, named voted-perceptron, is created to accumulate more information during training for generating superior predictions for the test data [138].

The characteristics of maintaining superior time complexity to deal with irrelevant features with perceptron-like linear algorithms has been discussed. This can be a substantial benefit for dealing with numerous features out of which only a few are pertinent ones. Usually, irrespective of the run time, all perceptron-like linear algorithms can be termed as anytime online algorithms which have the capacity to produce a useful outcome [139]. The longer they run, the better the result they produce. Finally, a multiclass problem must be reduced to a set of multiple binary classification problems as perceptron-like methods are binary.

2.9.3.2 Multilayer Perceptrons

Instances have to be linearly separable to make the learning arrive at a point where all instances have to be properly classified. This problem is addressed with multilayer perceptrons [140] (Fig. 2.6). A large number of units (neurons) joined together in a pattern of connections has constituted the multilayer neural network. Three major subdivisions of units are aimed at input units to receive information to be processed, output units to display the results of the processing and units in between that are better known as hidden units. Feed forward networks permit one-way movement of signals from input to output. The back propagation algorithm has divided the tasks into multiple steps. It provides the neural network with a training sample. The output of the network is compared to that of the expected output from the given sample to compute the error in each output neuron. The process is followed by calculation of desired output for each neuron along with a scaling factor for adjustment to match the desired output. This is known as local error. The weights of each neuron have to be adjusted to minimize the local error. Assignment of "blame" is to be done for the local error to the neurons at the previous level for handing over higher responsibilities to neurons connected

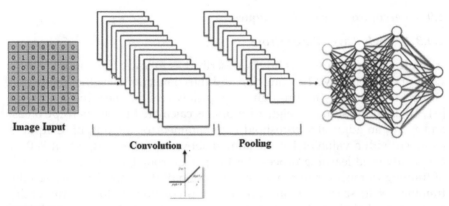

FIGURE 2.6
Perceptron-based Structure.

by stronger weights. The entire process is to be repeated on the neurons at previous levels by using the "blame" of each neuron as its error.

2.9.4 Statistical Learning Algorithm

An overt fundamental probability model has characterized the statistical approaches that define the likelihood for an instance to fit in each class, rather than a mere classification. Two simple methods, namely, linear discriminant analysis (LDA) and the related Fisher's linear discriminant, are exploited in statistics and machine learning to locate the linear combination of features that has the preeminent role to separate two or more classes of objects [141]. Measurements made on each observation have to be continuous quantities for LDA. A comparable technique for dealing with categorical variables has been discriminant correspondence analysis [142]. Another wide-ranging technique for estimating probability distributions from data is maximum entropy. The overriding principle followed by maximum entropy has to have maximum possible uniform distribution when nothing is known and have maximal entropy. Labeled training data is instrumental to develop a set of constraints for the model that portray the class-specific expectations for the distribution. A good tutorial preamble to maximum entropy techniques is presented in existing literature [143]. The most well-known representative of statistical learning algorithms is the Bayesian network. An illustrative presentation on Bayesian networks is covered in earlier approaches [144].

2.9.5 Support Vector Machine

The latest supervised machine learning technique is the Support Vector Machine (SVMs) [145]. SVMs encompass the concept of a "margin"—either side of a hyperplane that distinguishes two data classes. The reduction of an upper

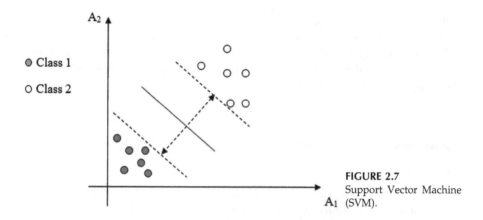

FIGURE 2.7
Support Vector Machine (SVM).

bound on the expected generalization error is observed by maximizing the margin, thereby creating the largest possible distance between the unraveling hyperplane and the instances on either side of it. An optimum separating hyperplane can be found by minimizing the squared norm of the separating hyperplane when linear separation of two classes can be done. Once the optimal untying hyperplane is identified, data points that lie on its margin are termed as support vector points and the solution has to be represented as a linear combination of only these points in case of linearly separable data by ignoring other data points. Thus, the model complexity of an SVM has remained unaltered by the number of features handled in the training data (the number of support vectors selected by the SVM learning algorithm is usually small) (Fig. 2.7). Therefore, learning tasks can be well handled where a large number of features have been encountered with respect to the number of training instances.

A broad variety of literature sources are available in the domain of content-based image classification. Different feature extraction techniques are carried out to extract a robust set of feature vectors from image data to enhance the classification accuracy. However, the contemporary classification techniques have the drawback of slow convergence due to hefty signature size. Assorted techniques for feature extraction and classification from image data have been surveyed in this chapter. It has presented elaborated insights of existing literature on content-based image classification.

Chapter Summary

This chapter has explored some of the prominent handcrafted techniques for extraction of content-based feature vectors from image data. It has discussed the recent advancements proposed by the research community for defining the content-based descriptors (features) based on various low-level features, namely, color, shape and texture. Furthermore, the chapter has elaborated on significant classification techniques based on decision trees, learning-based rules, statistical techniques, perceptron-based algorithms and support vector machine.

References

1. Lu, D. and Weng, Q., 2007. A survey of image classification methods and techniques for improving classification performance. *International Journal of Remote Sensing*, 28: 823–870.
2. Choras, R. S., 2007. Image feature extraction techniques and their applications for CBIR and biometrics systems. *International Journal of Biology and Biomedical Engineering*, 1(1): 6–16.
3. Singh, S. M. and Hemachandran, K., 2012. Content-based image retrieval using color moment and Gabor texture feature. *IJCSI International Journal of Computer Science*, 9(5): 299–309.
4. Datta, R., Joshi, D., Li, J. and Wang, J. Z., 2008. Image retrieval: Ideas, influences and trends of the New Age. *ACM Computing Surveys*, 40(2): 1–60.
5. Swain, M. J. and Ballard, D. H., 1991. Color indexing. *International Journal of Computer Vision*, 7(1): 11–32.
6. Deng, Y., Manjunath, B. S., Kenney, C., Moore, M. S. and Shin, H., 2001. An efficient color representation for image retrieval. *IEEE Transactions on Image Processing*, 10(1): 140–147.
7. Kotoulas, L. and Andreadis, I., 2003. Colour histogram content-based image retrieval and hardware implementation. *IEE Proceedings – Circuits, Devices and Systems*, 150(5): 387.
8. Han, J. and Ma, K. K., 2002. Fuzzy color histogram and its use in color image retrieval. *IEEE Transactions on Image Processing*, 11(8): 944–952.
9. Sezgin, M. and Sankur, B., 2004. Survey over image thresholding techniques and quantitative performance evaluation. *Journal of Electronic Imaging*, 13(1): 146.
10. Barney, S., Elisa H., Likforman-Sulem, L. and Darbon, J., 2010. Effect of pre-processing on binarization, eds. *Laurence Likforman-Sulem and Gady Agam*, Boise State University Scholar Works. 75340H-75340H-8.
11. Wang, Z., Li, S., Su, S. and Xie, G., 2009. Binarization algorithm of passport image based on global iterative threshold and local analysis. In 2009 International Conference on Information Management, Innovation Management and Industrial Engineering, IEEE, ICIII 2009, vol. 1, pp. 239–242.
12. Gatos, B., Pratikakis, I. and Perantonis, S. J., 2006. Adaptive degraded document image binarization. *Pattern Recognition*, 39(3): 317–327.
13. Kuo, T.-Y., Lai, Y. Y. and Lo, Y.-C., 2010. A novel image binarization method using hybrid thresholding. In Proceedings of ICME, IEEE, pp. 608–612.
14. Tong, L. J., Chen, K., Zhang, Y., Fu, X. L. and Duan, J. Y., 2009. Document image binarization based on NFCM. In Proceedings of the 2009 2nd International Congress on Image and Signal Processing, CISP'09, pp. 1–5.

15. Tanaka, H., 2009. Threshold correction of document image binarization for ruled-line extraction. In Proceedings of the International Conference on Document Analysis and Recognition, ICDAR, pp. 541–545.
16. Ntirogiannis, K., Gatos, B. and Pratikakis, I., 2008. An objective evaluation methodology for document image binarization techniques. 2008 The Eighth IAPR International Workshop on Document Analysis Systems, pp. 217–224.
17. Gatos, B., Pratikakis, I. and Perantonis, S. J., 2008. Improved document image binarization by using a combination of multiple binarization techniques and adapted edge information. 2008 19th International Conference on Pattern Recognition, pp. 1–4.
18. Pan, M. P. M., Zhang, F. Z. F. and Ling, H. L. H., 2007. An image binarization method based on HVS. *IEEE International Conference on Multimedia and Expo*, 2007, pp. 1283–1286.
19. Mello, C. A. B. and Costa, A. H. M., 2005. Image thresholding of historical documents using entropy and ROC curves. In *Progress in Pattern Recognition, Image Analysis and Applications*, eds. Manuel Lazo and Alberto Sanfeliu, Springer, vol. 3773, pp. 905–916.
20. Valizadeh, M., Armanfard, N., Komeili, M. and Kabir, E., 2009. A novel hybrid algorithm for binarization of badly illuminated document images. In 2009 14th International CSI Computer Conference, CSICC 2009, pp. 121–126.
21. Lu, S. and Chew, L. T., 2007. Binarization of badly illuminated document images through shading estimation and compensation. In Proceedings of the International Conference on Document Analysis and Recognition, ICDAR, vol. 1, pp. 312–316.
22. Chang, Y.-F., Pai, Y.-T. and Ruan, S.-J., 2009. An efficient thresholding algorithm for degraded document images based on intelligent block detection. IEEE International Conference on Systemic Man and Cybernetics. SMC, pp. 667–672.
23. Kekre, H. B., Thepade, S., Khandelwal, S., Dhamejani, K. and Azmi, A., 2011. Face recognition using multilevel block truncation coding. *International Journal of Computer Application*, 36(11): 38–44.
24. Kekre, H. B., Thepade, S., Khandelwal, S., Dhamejani, K. and Azmi, A., 2012. Improved face recognition using multilevel BTC using Kekre's LUV color space. *International Journal of Computer Application*, 3(1): 156–160.
25. Shaikh, S. H., Maiti, A. K. and Chaki, N., 2013. A new image binarization method using iterative partitioning. *Machine Vision and Applications*, 24(2): 337–350.
26. Otsu, N., 1979. A threshold selection method from gray-level histogram. *IEEE Transactions on Systems, Man, and Cybernetics*, 9: 62–66.
27. Johannsen, G. and Bille, J., 1982. A threshold selection method using information measures. In 6th International Conference on Pattern Recognition, pp. 140–143.
28. Liu C., 2013. A new finger vein feature extraction algorithm. In IEEE 6th International Congress on Image and Signal Processing, CISP, pp. 395–399.
29. Ramírez-Ortegón, M. A. and Rojas, R., 2010. Unsupervised evaluation methods based on local gray-intensity variances for binarization of historical documents. In Proceedings – International Conference on Pattern Recognition, pp. 2029–2032.
30. Yanli, Y. and Zhenxing, Z., 2012. A novel local threshold binarization method for QR image, In IET International Conference on Automatic Control and Artificial Intelligence, ACAI, pp. 224–227.

31. Bernsen, J., 1986. Dynamic thresholding of gray level images. In Proceedings of International Conference on Pattern Recognition, ICPR, pp. 1251–1255.
32. Kittler, J. and Illingworth, J., 1986. Minimum error thresholding. *Pattern Recognition*, 19(1): 41–47.
33. Annadurai, S. and Shanmugalakshmi, R., 2011. Image transforms. In *Fumdamentals of Digital Image Processing*, Pearson, pp. 31–66.
34. Kekre, H. B. and Thepade, S., 2009. Improving the performance of image retrieval using partial coefficients of transformed image. *International Journal of Information Retrieval, Serials Publications*, 2(1): 72–79.
35. Kekre, H. B., Thepade, S. and Maloo, A., 2010. Image retrieval using fractional coefficients of transformed image using DCT and Walsh transform. *International Journal of Engineering Science and Technology, IJEST*, 2(4): 362–371.
36. Kekre, H. B., Thepade, S. and Maloo, A., 2010. Performance comparison of image retrieval using row mean of transformed column image. *International Journal on R-20 Computer Science and Engineering, IJCSE*, 2(5): 959–964.
37. Kekre, H. B., Thepade, S. and Maloo, A., 2010. Image retrieval using row means of column transformed even and odd parts of image with Walsh, Haar and Kekre transforms. *International Journal of Computer Science and Information Security (IJCSIS)*, 8(5): 360–363.
38. Kekre, H. B., Thepade, S., Athawale, A., Anant, S., Prathamesh, V. and Suraj, S., 2010. Kekre transform over row mean, column mean and both using image tiling for image retrieval. *International Journal of Computer and Electrical Engineering, IJCEE*, 2(6): 964–971.
39. Kekre, H. B., Thepade, S. and Maloo, A., 2010. CBIR feature vector dimension reduction with eigenvectors of covariance matrix using row, column and diagonal mean sequences. *International Journal of Computer Applications, IJCA*, 3(12): 786–1114.
40. Prakash, O., Khare, M., Srivastava, R. K. and Khare, A., 2013. Multiclass image classification using multiscale biorthogonal wavelet transform. In IEEE Second International Conference on Information Processing, ICIIP, pp. 131–135.
41. Luo, H., Yang, L, Yuan, H. and Tang, Y. Y., 2013. Dimension reduction with randomized anisotropic transform for hyperspectral image classification. In 2013 IEEE International Conference on Cybernetics, CYBCONF 2013, pp. 156–161.
42. Rajan, A., Ramesh, G. P. and Yuvaraj, J., 2014. Glaucomatous image classification using wavelet transform. In IEEE International Conference on Advanced Communication Control and Computing Technologies, ICACCCT, 2014, pp. 1398–1402.
43. Wang, Y. and Cui, S., 2014. Hyperspectral image feature classification using stationary wavelet transform. In IEEE International Conference on Wavelet Analysis and Pattern Recognition, ICWAPR, 2014, pp. 104–108.
44. Suma, K. V., Indira, K. and Rao, B., 2014. Classification of nailfold capillary images using wavelet and Discrete Cosine Transform. In IEEE International Conference on Circuits, Communication, Control and Computing, I4C, 2014, pp. 105–108.
45. Kobayashi, T., 2014. Dirichlet-based histogram feature transform for image classification. In IEEE Conference on Computer Vision and Pattern Recognition, CVPR, 2014: 3278–3295.

46. Rezaeilouyeh, H., Mahoor, M. H., Mavadati, S. M., Zhang, J. J., 2013. A microscopic image classification method using shearlet transform. In IEEE International Conference on Healthcare Informatics, ICHI, 2013, pp. 382–386.

47. Li, Y. H. and Savvides M., 2007. Kernel fukunaga-koontz transform subspaces for enhanced face recognition. In Proceedings of the IEEE Computer Society Conference on Computer Vision and Pattern Recognition, pp. 1–8.

48. Hafed, Z. M. and Levine, M. D., 2001. Face recognition using the discrete cosine transform. *International Journal of Computer Vision*, 43(3): 167–188.

49. Shen, G. L. and Wu, X. J., 2013. Content based image retrieval by combining color, texture and CENTRIST. In IEEE Constantinides International Workshop on Signal Processing, CIWSP, 2013, pp. 1–4.

50. Walia, E., Goyal, A. and Brar, Y.S., 2013. Zernike moments and LDP-weighted patches for content-based image retrieval. *Signal, Image and Video Processing, Springer*, 8(3): 577–594.

51. Kekre, H. B., Thepade, S., Wadhwa, S., Singh, S., Kakaiya, M. and Mukherjee, P., 2010. Image retrieval with shape features extracted using gradient operators and slope magnitude technique with BTC. *International Journal of Computer Applications, IJCA*, 6(8): 28–33.

52. Mehrotra, R. and Gary, J. E., 1995. Similar-shape retrieval in shape data management. *IEEE Computer*, 28(9): 57–62.

53. Berretti, S., Del Bimbo, A., and Pala, P., 2000. Retrieval by shape similarity with perceptual distance and effective indexing. *IEEE Transaction on Multimedia*, 2(4): 225–239.

54. Petrakis, E. G. M., Diplaros, A. and Milios, E., 2002. Matching and retrieval of distorted and occluded shapes using dynamic programming. *IEEE Transaction on Pattern Analysis and Machine Intelligence*, 24(4): 509–522.

55. Flickner, M., Sawhney, H., Niblack, W., Ashley, J., Huang, Q., Dom, B., Gorkani, M., et al., 1995. Query by image and video content: the QBIC system. *Computer*, 28(9): 23–32.

56. Latecki L. J. and Lakamper R., 2000. Shape similarity measure based on correspondence of visual parts. *IEEE Transaction on Pattern Analysis Machine Intelligence*, 22(10): 1185–1190.

57. Belongie S., Malik J. and Puzicha J., 2002. Shape matching and object recognition using shape contexts. *IEEE Transactions Pattern Analysis and Machine Intelligence*, 24(4): 509–522.

58. Ciaccia, P., Patella, M., and Zezula, P., 1997. M-tree: An efficient access method for similarity search in metric spaces. In Proceedings of the 23rd International Conference on Very Large Data Bases, Morgan Kaufmann, pp. 426–435.

59. Benediktsson, J. A., Palmason, J. A., and Sveinsson, J. R., 2005. Classification of hyperspectral data from urban areas based on extended morphological profiles. *IEEE Transactions on Geoscience and Remote Sensing*, 43(3): 480–491.

60. Fesharaki, N. J. and Pourghassem, H., 2012. Medical X-ray images classification based on shape features and Bayesian rule. In 2012 Fourth International Conference on Computational Intelligence and Communication Networks, IEEE, 369–373.

61. Ganapathi, K., Madumbu, V., Rajendran, R. and David, S., with Phalguni, 2013. Design and implementation of an automatic traffic sign recognition system on

TI OMAP-L138. In Proceedings of the IEEE International Conference on Industrial Technology, pp. 1104–1109.

62. Parape, C. D. K., Premachandra, H. C. N., Tamura, M., Sugiura, M., 2012. Damaged building identifying from VHR satellite imagery using morphological operators in 2011 Pacific coast of Tohoku Earthquake and Tsunami. In IEEE International Geoscience and Remote Sensing Symposium, IGARSS, 2012, pp. 3006–3009.

63. Na, W. and Tao, W., 2012. Proximal support vector machine based pavement image classification. In IEEE Fifth International Conference on Advanced Computational Intelligence, ICACI, 2012, pp. 686–688.

64. Wang, Z., Lu, B., Chi, Z. and Feng, D., 2011. Leaf image classification with shape context and SIFT descriptors. In Proceedings – 2011 International Conference on Digital Image Computing: Techniques and Applications, DICTA, 2011, pp. 650–654.

65. Lucchi, A., Smith, K., Achanta, R., Knott, G. and Fua, P., 2012. Supervoxel-based segmentation of mitochondria in em image stacks with learned shape features. *IEEE Transactions on Medical Imaging*, 31(2): 474–486.

66. Amanatiadis, A., Kaburlasos, V. G., Gasteratos, A. and Papadakis, S. E., 2011. Evaluation of shape descriptors for shape-based image retrieval. *IET Image Processing*, pp. 493–499.

67. Wang, X., Feng, B., Bai, X., Liu, W. and Latecki, L. J., 2014. Bag of contour fragments for robust shape classification. *Pattern Recognition*, 47(6): 2116–2125.

68. Sun, K. B. and Super, B. J., 2005. Classification of contour shapes using class segment sets. In *Proceedings of the IEEE Computer Society Conference on Computer Vision and Pattern Recognition*, vol. 2, pp. 727–733.

69. Shen, W., Bai, X., Hu, R., Wang, H. and Latecki, L. J., 2011. Skeleton growing and pruning with bending potential ratio. *Pattern Recognition*, 44(2): 196–209.

70. Daliri, M. R. and Torre, V., 2008. Robust symbolic representation for shape recognition and retrieval. *Pattern Recognition*, 41(5): 1799–1815.

71. Bai, X., Wang, X., Latecki, L. J., Liu, W. and Tu, Z., 2009. Active skeleton for non-rigid object detection. In Proceedings of the IEEE International Conference on Computer Vision, pp. 575–582.

72. Demirci, M. F., Shokoufandeh, A. and Dickinson, S. J., 2009. Skeletal shape abstraction from examples. *IEEE Transactions on Pattern Analysis and Machine Intelligence*, 31(5): 944–952.

73. Yu, X., Yi, L., Fermuller, C. and Doermann, D., 2007. Object detection using a shape codebook. In British Machine Vision Conference, vol. 4. http://www.comp.leeds.ac.uk/bmvc2008/proceedings/2007/papers/paper-13.pdf.

74. Sundar, H., Silver, D., Gagvani, N. and Dickinson, S., 2003. Skeleton based shape matching and retrieval. In Proceedings – SMI 2003: Shape Modeling International 2003, pp. 130–139.

75. Bai, X., Liu, W. and Tu, Z., 2009. Integrating contour and skeleton for shape classification. In 2009 IEEE 12th International Conference on Computer Vision Workshops, ICCV Workshops 2009, pp. 360–367.

76. Bai, X. and Latecki, L. J., 2008. Path similarity skeleton graph matching. *IEEE Transactions on Pattern Analysis and Machine Intelligence*, 30(7): 1282–1292.

77. Belongie, S., Belongie, S., Malik, J., Malik, J., Puzicha, J. and Puzicha, J., 2002. Shape matching and object recognition using shape contexts. *IEEE Transactions on Pattern Analysis and Machine Intelligence*, 24: 509–522.

78. Mokhtarian, F. and Suomela, R., 1998. Robust image corner detection through curvature scale space. *IEEE Transactions on Pattern Analysis and Machine Intelligence*, 20(12): 1376–1381.

79. Alajlan, N., Rube, I. E., Kamel, M. S. and Freeman, G., 2007. Shape retrieval using triangle-area representation and dynamic space warping. *Pattern Recognition*, 40(7): 1911–1920.

80. McNeill, G., and Vijayakumar, S., 2006. Hierarchical procrustes matching for shape retrieval. In Proceedings of the IEEE Computer Society Conference on Computer Vision and Pattern Recognition, vol. 1, pp. 885–892.

81. Keshet, R., 2007. Adjacency lattices and shape-tree semilattices. *Image and Vision Computing*, 25(4): 436–446. SPEC. ISS.

82. Xu, C., Liu, J. and Tang, X., 2009. 2D shape matching by contour flexibility. *IEEE Transactions on Pattern Analysis and Machine Intelligence*, 31(1): 180–186.

83. Thayananthan, A., Stenger, B., Torr, P. H. S. and Cipolla, R., 2003. Shape context and chamfer matching in cluttered scenes. In Proceedings of the IEEE Computer Society Conference on Computer Vision and Pattern Recognition, pp. I–I.

84. Ling, H. and Jacobs, D. W., 2007. Shape classification using the inner-distance. *IEEE Transactions on Pattern Analysis and Machine Intelligence*, 29(2): 286–299.

85. Randen, T. and Husøy, J. H., 1999. Filtering for texture classification: A comparative study. *IEEE Transactions on Pattern Analysis and Machine Intelligence*, 21(4): 291–310.

86. Lin, C.-H., Liu, C.-W. and Chen, H.-Y., 2012. Image retrieval and classification using adaptive local binary patterns based on texture features. *Image Processing, IET*, 6(7): 822–830.

87. Nanthagopal, A. P. and Sukanesh, R., 2013. Wavelet statistical texture features-based segmentation and classification of brain computed tomography images. *Image Processing, IET*, 7(1): 25–32.

88. Sun, Z., Zhang, H., Tan, T. and Wang, J., 2014. Iris image classification based on hierarchical visual codebook. *IEEE Transactions on Pattern Analysis and Machine Intelligence*, 36(6): 1120–1133.

89. Regniers, O., Da Costa, J.-P., Grenier, G., Germain, C. and Bombrun, L., 2013. Texture based image retrieval and classification of very high resolution maritime pine forest images. *Geoscience and Remote IEEE International Sensing Symposium, IGARSS*, 2013: 4038–4041.

90. Kekre, H. B., Bharadi, V. A., Thepade, S. D., Mishra, B. K., Ghosalkar, S. E. and Sawant, S. M., 2010. Content based image retrieval using fusion of Gabor magnitude and Modified Block Truncation Coding. In Proceedings – 3rd International Conference on Emerging Trends in Engineering and Technology, ICETET 2010, pp. 140–145.

91. Vo, A., and Oraintara, S., 2008. Using phase and magnitude information of the complex directional filter bank for texture segmentation. In European Signal Processing Conference, pp. 1–5.

92. Haralick R., 1979. Statistical and structural approaches to texture. *Proceedings of IEEE*, 67(5): 786–804.

93. Malik, J. and Perona, P., 1990. Preattentive texture discrimination with early vision mechanisms. *Journal of the Optical Society of America. A, Optics and image science*, 7(5): 923–932.

94. Ma, W. Y. and Manjunath, B. S., 1995. A comparison of wavelet transform features for texture image annotation. In Proceedings of the International Conference on Image Processing, pp. 256–259.

95. Laine, A. and Fan, J., 1993. Texture classification by wavelet packet signatures. *IEEE Transactions on Pattern Analysis and Machine Intelligence*, 15(11): 1186–1191.

96. Thyagarajan, K., Nguyen, T. and Persons, C. E., 1994. A maximum likelihood approach to texture classification using wavelet transform, In IEEE International Conference Image Processing, 1994. Proceedings. ICIP-94, vol. 2, pp. 640–644.

97. Do, M. N. and Vetterli, M., 2002. Wavelet-based texture retrieval using generalized Gaussian density and Kullback–Leibler distance. *IEEE Transactions on Image Processing*, 11(2): 146–158.

98. Li, J., Gray, R. M. and Olshen, R. A., 2000. Multiresolution image classification by hierarchical modeling with two-dimensional hidden Markov models. *IEEE Transactions on Information Theory*, 46(5): 1826–1841.

99. Portilla, J. and Simoncelli, E. P., 2000. Parametric texture model based on joint statistics of complex wavelet coefficients. *International Journal of Computer Vision*, 40(1): 49–71.

100. Ma, W. Y. and Manjunath, B. S., 1996. *A Pattern Thesaurus For Browsing Large Aerial Photographs*. Technical Report 96-10, University of California at Santa Barbara.

101. Ma, W. Y. and Manjunath, B. S., 1998. A texture thesaurus for browsing large aerial photographs. *Journal of the American Society for Information Science*, 49(7): 633–648.

102. Manjunath, B.S., Ohm, J.R., Vasudevan, V.V. and Yamada, A., 2001. Color and texture descriptors. *IEEE Transactions on Circuits and Systems for Video Technology*, 11(6): 703–715.

103. Ma, W. Y. and Manjunath, B. S., 1996. Texture features and learning similarity. In IEEE Computer Society Conference on Computer Vision and Pattern Recognition, pp. 425–430.

104. Jain, A. K. and Farrokhnia, F., 1991. Unsupervised texture segmentation using Gabor filters. *Pattern Recognition*, 24(12): 1167–1186.

105. Schaffalitzky, F. and Zisserman, A., 2001. Viewpoint invariant texture matching and wide baseline stereo. In Proceedings Eighth IEEE International Conference on Computer Vision. ICCV 2001, pp. 636–643.

106. Mikolajczyk, K. and Schmid, C., 2004. Scale & affine invariant interest point detectors. *International Journal of Computer Vision*, 60(1): 63–86.

107. Hiremath, P. S. and Pujari, J., 2007. Content based image retrieval using color, texture and shape features. *15th International Conference on Advanced Computing and Communications ADCOM 2007*, 9(2): 780–784.

108. Yue, J., Li, Z., Liu, L. and Fu, Z., 2011. Content-based image retrieval using color and texture fused features. *Mathematical and Computer Modelling*, 54(3–4): 1121–1127.

109. Banerjee, M., Kundu, M. K. and Maji, P., 2009. Content-based image retrieval using visually significant point features. *Fuzzy Sets and Systems*, 160(23): 3323–3341.

110. Jalab, H. A., 2011. Image retrieval system based on color layout descriptor and Gabor filters. 2011 IEEE Conference on Open Systems, pp. 32–36.

111. Irtaza, A., Jaffar, M. A., Aleisa, E. and Choi, T. S., 2013. Embedding neural networks for semantic association in content based image retrieval. *Multimedia Tool and Applications*, 72(2): 1–21.

112. Rahimi, M. and Moghaddam, M. E., 2013. A content based image retrieval system based on color ton distributed descriptors. *Signal Image and Video Processing*, 9(3): 1–14.

113. Elalami, M. E., 2014. A new matching strategy for content based image retrieval system. *Applied Soft Computing Journal*, 14(C): 407–418.

114. Subrahmanyam, M., Jonathan Wu, Q. M., Maheshwari, R. P. and Balasubramanian, R., 2013. Modified color motif co-occurrence matrix for image indexing and retrieval. *Computers and Electrical Engineering*, 39(3): 762–774.

115. Walia, E., Vesal, S. and Pal, A., 2014. An effective and fast hybrid framework for color image retrieval. *Sensing and Imaging, Springer*, 15(1): 1–23.

116. Subrahmanyam, M., Maheshwari, R. P. and Balasubramanian, R., 2012. Expert system design using wavelet and color vocabulary trees for image retrieval. *Expert Systems with Applications*, 39(5): 5104–5114.

117. Das, R. and Walia, E., 2019. Partition selection with sparse autoencoders for content based image classification. *Neural Computing and Applications*, 31(3): 675–690.

118. Phyu, T. N., 2009, March. Survey of classification techniques in data mining. In Proceedings of the International MultiConference of Engineers and Computer Scientists, vol. 1, pp. 18–20.

119. Ture, M., Kurt, I., Kurum, A. T. and Ozdamar, K., 2005. Comparing classification techniques for predicting essential hypertension. *Expert Systems with Applications*, 29(3): 583–588.

120. Lindsey, D. T., 2000. Vision science: Photons to phenomenology. *Optometry and Vision Science*, 77(5): 233–234.

121. Forsyth, D. A. and Ponce, J., 2002. Computer vision: a modern approach. *Prentice Hall Professional Technical Reference*.

122. Hunt E., Martin J. and Stone P., 1966. *Experiments in Induction*, New York, Academic Press.

123. Loh, W. Y., 2011. Classification and regression trees. *Wiley Interdisciplinary Reviews: Data Mining and Knowledge Discovery*, 1(1): 14–23.

124. Kononenko, I., 1991. Semi-naive Bayesian classifier. In Proceedings of the Sixth European Working Session on Learning, pp. 206–219.

125. Murthy, 1998 Automatic construction of decision trees from data: A multidisciplinary survey. *Data Mining and Knowledge Discovery*, 2: 345–389.

126. Zheng, Z., 1998. Constructing conjunctions using systematic search on decision trees. *Knowledge Based Systems Journal*, 10: 421–430.

127. Zheng, Z., 2000. Constructing X-of-N attributes for decision tree learning. *Machine Learning*, 40: 35–75.

128. Gama, J. and Brazdil, P., 1999. Linear tree. *Intelligent Data Analysis*, 3: 1–22.

129. Quinlan, J. R., 1993. *C4.5: Programs for Machine Learning*. Morgan Kaufmann, San Francisco.

130. Fürnkranz, J., 1999. Separate-and-conquer rule learning. *Artificial Intelligence Review*, 13(1): 3–54.

131. Smyth, P. and Goodman, R. M., 1991. Rule induction using information theory. In *Knowledge Discovery in Databases, KDD*, pp. 159–176.

132. An, A. and Cercone, N., 2000. Rule quality measures improve the accuracy of rule induction: An experimental approach. *In Foundations of Intelligent Systems,* 1932: 119–129.

133. Furnkranz, J., Flach, P., 2005. ROC 'n' rule learning—towards a better understanding of covering algorithms. *Machine Learning,* 58(1), pp. 39–77.

134. Cohen, W. W., 1995. Fast effective rule induction. In Proceedings of the Twelfth International Conference on Machine Learning, pp. 115–123.

135. An, A. and Cercone, N., 1999. Discretization of continuous attributes for learning classification rules. Third Pacific-Asia Conference on Methodologies for Knowledge Discovery & Data Mining, pp. 509–514.

136. Flach, P. A. and Lavrac, N., 2000, July. The role of feature construction in inductive rule learning. In Proceedings of the ICML2000 Workshop on Attribute-Value and Relational Learning: Crossing the Boundaries, pp. 1–11.

137. Markovitch, S. and Rosenstein, D., 2002. Feature generation using general constructor functions. *Machine Learning,* 49(1): 59–98.

138. Freund, Y. and Schapire, R. E., 1999. Large margin classification using the perceptron algorithm. *Machine Learning,* 37(3): 277–296.

139. Kivinen, J., 2002. Online learning of linear classifiers, *Advanced Lectures on Machine Learning: Machine Learning Summer School 2002, Australia,* pp. 235–257. February 11–22, ISSN: 0302-9743.

140. Rumelhart, D. E., Hinton, G. E. and Williams, R. J., 1986. Learning internal representations by error propagation. *In Parallel Distributed Processing: Explorations in the Microstructure of Cognition,* 1: 318–362.

141. Friedman, J. H., 1989. Regularized discriminant analysis. *Journal of the American Statistical Association,* 84(405): 165–175.

142. Mika, S., Ratsch, G., Weston, J. and Scholkopf, B., 1999. Fisher discriminant analysis with kernels. Neural Networks for Signal Processing IX, 1999. In Proceedings of the 1999 IEEE Signal Processing Society Workshop, pp. 41–48.

143. Csiszár, I., 1996. Maxent, mathematics, and information theory. In Proceedings of the Fifteenth International Workshop on Maximum Entropy and Bayesian Methods, Santa Fe, New Mexico, USA, 1995, vol. 79, pp. 35–50.

144. Jensen, F., 1996. *An Introduction to Bayesian Networks.* Springer.

145. Vapnik, V. N., 1995. *The Nature of Statistical Learning Theory.* Springer, vol. 8.

3

Content-Based Feature Extraction: Color Averaging

3.1 Prelude

One of the significant primary attributes of any image is its color, which can be utilized to generate meaningful features for content-based image classification. Rich literature is available on the proposition of assorted techniques for generating robust descriptors with color features. A popular algorithm, color difference histogram (CDH), extracts color features by computing intuitively the consistent color difference between two points with diverse backgrounds [1,2].

Efficacy of image indexing and recognition with color is well formulated by using color correlogram by means of joint probability [3]. Color distribution is considered a universal image characteristic to stimulate content-based image recognition in recent literature [4]. Another efficient technique of designing descriptors is color histogram, which largely has been studied and implemented in recent works [5,6].

Color moments (CM) have also shown significant performance as a robust feature vector to enhance precision for content-based image classification [5].

The state-of-the-art techniques of extracting feature vectors using color features have motivated the proposition and implementation of two novel techniques for extraction of feature vectors using color characteristics of images. The following techniques have explored the color features to generate meaningful descriptors for content-based image classification.

- Block truncation coding (BTC) with color clumps
- Sorted block truncation coding (SBTC) for content-based image classification

Prior to elaborating on the algorithms behind the two techniques, the following section briefly discusses the concept of block truncation coding.

3.2 Block Truncation Coding

Block truncation coding (BTC) is one of the widely accepted compression algorithms due to its simplicity of implementation. The working principle of this algorithm initiates with segmenting the image into $n \times n$ (typically 4×4) non-overlapping blocks [7]. Primarily developed for grayscale images in 1979, the algorithm was one of the benchmarked techniques during the inception of image processing. The popularity of the algorithm motivated developers to extend it for color images at a later stage. The algorithm functions by coding the image blocks one at a time. Fresh calculation of values for the reconstructed blocks is based on the mean and standard deviation for each block, which maintains values of mean and standard deviation similar to that of the original block (Fig. 3.1).

3.3 Feature Extraction Using Block Truncation Coding with Color Clumps

The concept of BTC is used as a precursor of this feature extraction technique by differentiating each color component—namely, Red (R), Green (G) and Blue

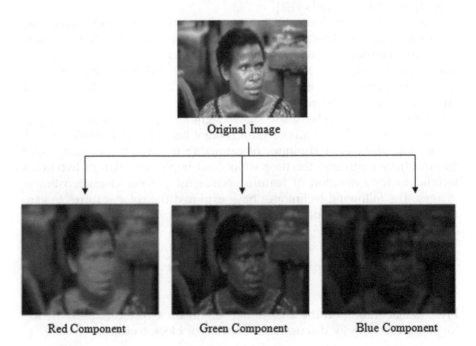

Original Image

Red Component   Green Component   Blue Component

FIGURE 3.1
Block Truncation Coding with R, G and B Components as Blocks.

(B)—as blocks from the images. Henceforth, the computation of color clumps for feature extraction is carried out with the Wang dataset in RGB color space. A step-by-step guide for feature extraction using color clumps follows:

- Color clump calculation initiates with approximation of gray values for individual color components to form disjoint subsets of gray values for the corresponding components.
- Next is the logical partition formation of the intensity values into numerous subdivisions of color clumps, which involves several stages of color bin formation for each R, G and B color component.
- Increase in number of partitions is formulated in multiples of 2 as described in Fig. 3.4.
- The partitions are evaluated in the range of 2 to 32 to identify the suitable number of clumps for feature vector extraction.
- Each of the five stages of feature extraction is named as BTC with 2 color clumps, BTC with 4 color clumps, BTC with 8 color clumps, BTC with 16 color clumps and BTC with 32 color clumps respectively.
- All the gray values of a particular partition are compared to that of the mean of the gray values for the corresponding partition for each color clump.
- The process eventually creates two different intensity groups having values greater than the mean threshold and lesser than the mean threshold respectively.
- The mean upper intensity and mean lower intensity values for the corresponding clumps are calculated.

Thus, one can compute two different feature values of higher and lower intensities from a particular clump. This makes the number of feature vectors double the number of the color clumps selected (Fig. 3.2).

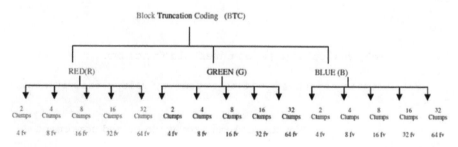

FIGURE 3.2
Distribution of Color Clumps and Feature Vectors (fv) for Each Color.

3.4 Code Example (MATLAB®)

```
#Read an image
i=imread('path\imagefile');
#Separate the color components
i
1=i(:,:,1); #red component
[m n p]=size(i1); #dimension of red component
i3=i(:,:,2); #green component
[x y z]=size(i3); #dimension of green component
i6=i(:,:,3); #blue component
[j k l]=size(i6); #dimension of blue component

#Computing the means of different color clumps for red component
for a1 = 1:m #for loop for row/column iteration of red component
for a2 = 1:n
if i1(a1,a2)>=0 && i1(a1,a2)<32 #specifying the first clump range
c1=c1+1;
s1=s1+i1(a1,a2); #summing up intensity values of first clump
end
end
end
if c1~=0
TavR1 = s1/c1; #calculating mean threshold for clump
else
TavR1=0; #mean threshold 0 if no intensity values fall in clump
range
end
for a3 = 1:m #for loop for row/column iteration of red component
for a4 = 1:n
if i1(a3,a4)>=32 && i1(a3,a4)< 64 #specifying the second clump
range
c2=c2+1; #summing up intensity values of second clump
s2=s2+i1(a3,a4);
end
end
end
if c2~=0
TavR2 = s2/c2; #calculating mean threshold for clump
else
TavR2 = 0; #mean threshold 0 if no intensity values fall in clump
range
end
for a5 = 1:m #for loop for row/column iteration of red component
for a6 = 1:n
if i1(a5,a6)>=64 && i1(a5,a6)< 96 #specifying the third clump range
c3=c3+1;
```

```
s3=s3+i1(a5,a6); #summing up intensity values of third clump
end
end
end
if c3~=0
TavR3 = s3/c3; #calculating mean threshold for clump
else
TavR3 = 0; #mean threshold 0 if no intensity values fall in clump
range
end
for a7 = 1:m #for loop for row/column iteration of red component
for a8 = 1:n
if i1(a7,a8)>=96 && i1(a7,a8)< 128 #specifying the fourth clump
range
c4=c4+1;
s4=s4+i1(a7,a8); #summing up intensity values of fourth clump
end
end
end
if c4~=0
TavR4 = s4/c4; #calculating mean threshold for clump
else
TavR4=0; #mean threshold 0 if no intensity values fall in clump range
end

#Feature Vector Extraction from color clumps
for a25=1:m
for a26=1:n
if i1(a25,a26)>=TavR1 #finding intensity values higher than mean
threshold
c13=c13+1;
s13=s13+R(a25,a26); #summing up high-intensity values
end
end
end
if c13~=0
upR1=s13/c13; #calculating mean high intensity as feature vector
else
upR1=0;
end
for a27=1:m
for a28=1:n
if i1(a27,a28)<TavR1 #finding intensity values lower than mean
threshold
c14=c14+1;
s14=s14+R(a27,a28); #summing up low-intensity values
end
end
```

```
end
if c14~=0
loR1=s14/c14; #calculating mean low intensity as feature vector
else
loR1=0;
end
```

3.5 Coding Exercise

The previous code example is used for designing 16 clumps for red color component. Refer to the example to complete the following assignments:

- Compute the mean by designing clumps for green and blue color components
- Generate feature vectors of dimension 4, 8, 16, 32 and 64

Comparative results of classification performances have revealed maximum Precision and Recall values with 16 clumps for the mentioned approach. This has provided a feature vector of dimension 32 for each color component, which sums up to 96 for formation of feature vector with all three color components irrespective of image dimension.

Initial comparison of results in Table 3.1 embraces the metrics named Misclassification Rate (MR) and F1-Score. The number of color clumps is denoted by N, and the Wang dataset is considered a testbed for classification using the KNN Classifier.

The observations for the experiments follow:

- MR decreases and F1-Score increases till $N = 16$.
- Lowest MR recorded is 0.050 for $N = 16$, and the subsequent highest F1-Score is 0.728.
- Two observations indicate the common inference of higher classification performance until $N = 16$ subsets of the intensity values.
- Further enhancement of N for $N = 32$ has decreased F1-Score to 0.56 and amplified MR to 0.099.
- Therefore, the number of clumps to be designed is restricted to 16 for the work.

Hereafter, calculation of average values for diverse metrics is carried out to observe Precision, Recall, MR and F1-Score for classification with four different classifiers—namely, KNN, Ripple-Down Rule (RIDOR), Artificial Neural Network (ANN) and Support Vector Machine (SVM)—for partition $N = 16$ used for feature extraction with color clumps from four different public datasets.

TABLE 3.1

Misclassification Rate (MR) and the F1-Score Comparison for Different Values of Color Clumps Subset (*N*) While Classifying with KNN Classifier

Categories	Color Clumps $N = 2$ MR	Color Clumps $N = 2$ F1-Score	Color Clumps $N = 4$ MR	Color Clumps $N = 4$ F1-Score	Color Clumps $N = 8$ MR	Color Clumps $N = 8$ F1-Score	Color Clumps $N = 16$ MR	Color Clumps $N = 16$ F1-Score	Color Clumps $N = 32$ MR	Color Clumps $N = 32$ F1-Score
Tribals	0.08	0.65	0.08	0.69	0.07	0.73	0.06	0.75	0.12	0.42
Sea Beach	0.08	0.6	0.08	0.61	0.08	0.56	0.08	0.59	0.16	0.4
Gothic Structure	0.13	0.42	0.14	0.46	0.14	0.44	0.13	0.48	0.17	0.13
Bus	0.1	0.54	0.07	0.68	0.08	0.66	0.05	0.78	0.12	0.45
Dinosaur	0	1	0	1	0	0.99	0	1	0	1
Elephant	0.06	0.72	0.05	0.76	0.05	0.79	0.07	0.7	0.09	0.58
Roses	0.02	0.91	0.01	0.94	0.02	0.91	0.02	0.93	0.03	0.85
Horses	0.02	0.91	0.02	0.92	0.01	0.96	0.02	0.9	0.04	0.81
Mountains	0.13	0.37	0.1	0.47	0.1	0.51	0.11	0.43	0.15	0.38
Average	0.07	0.681	0.06	0.725	0.06	0.727	0.059	0.728	0.099	0.56

[Observation: $N = 16$ has the highest F1-Score and lowest MR for KNN Classifier.]

TABLE 3.2

Classification Performance by Feature Extraction Using BTC with Color Clumps
with Four Different Public Datasets Using Four Different Classifiers

Datasets	Metrics	KNN	RIDOR	ANN	SVM
Wang	Precision	0.741	0.654	0.914	0.715
	Recall	0.737	0.657	0.913	0.727
	MR	0.059	0.079	0.038	0.066
	F1-Score	0.728	0.655	0.913	0.717
OT-Scene	Precision	0.461	0.368	0.484	0.501
	Recall	0.45	0.36	0.479	0.494
	MR	0.096	0.1	0.091	0.09
	F1-Score	0.446	0.361	0.481	0.495
Corel	Precision	0.596	0.512	0.656	0.653
	Recall	0.581	0.519	0.664	0.65
	MR	0.08	0.083	0.078	0.079
	F1-Score	0.584	0.515	0.658	0.637
Caltech	Precision	0.516	0.491	0.565	0.475
	Recall	0.553	0.489	0.574	0.557
	MR	0.091	0.09	0.081	0.09
	F1-Score	0.484	0.489	0.568	0.484

The comparative results are given in Table 3.2.

3.6 Feature Extraction Using Sorted Block Truncation Coding for Content-Based Image Classification

Image processing algorithms have considered noise extending adverse influence on image data by staining it to a large extent. The fact is evident in
plenty of literature that has introduced varied noise elimination techniques
to decrease the effect of noise over image data. One of the popular techniques is the binning method, which is extensively implemented for
smoothing sorted values of data in a dataset [8]. A stepwise functionality of
binning method to sort the values in a dataset follows:

- The sorted values are divided in bins.
- Average of the sorted values are computed by considering the
  neighboring values to perform local smoothing for each of the bins.
  The local smoothing is crucial for noise removal.
- The corresponding average value for a particular bin is instru-
  mental in replacing each value present in the bin.

- The process is repeated for all the values in the bin to perform smoothing.

A similar approach is adopted in the case of SBTC by implementing modified binning technique. The gray values of images are sorted for feature vector extraction from individual color components Red (R), Green (G) and Blue (B).

The primary difference of this technique with the one described in Section 3.3 lies in the dimension of feature vector extracted. The earlier technique creates feature vectors having double dimension to the number of bins, whereas, this technique results in feature vectors equal to the number of bins as in Fig. 3.3.

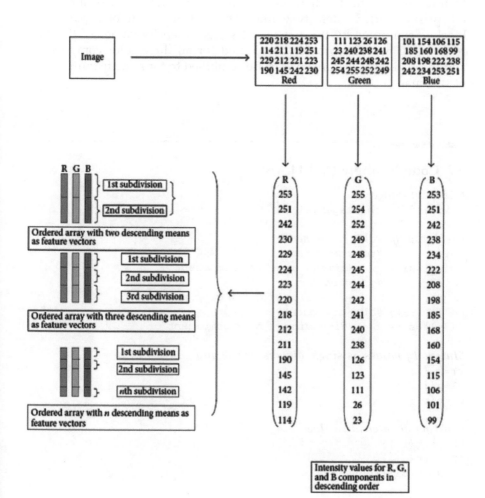

FIGURE 3.3
Block Diagram for SBTC.

The intensity values of an image are represented by the process with an array of single dimension. Sorting of the single-dimension array is carried out in ascending order.

For instance, sorting and dividing all the gray values into $N = 8$ bins or partitions will result in the generation of 8 feature vectors, which turns out to be identical to the number of bins or partitions. This results in comprehensive reduction in the size of the feature vector, thus enhancing the efficiency of computation for extraction of feature vectors. The process has separated the intensity values into N blocks where ($N \geq 2$). Block truncation coding technique has considered each of the color components Red (R), Green (G) and Blue (B) as a block to implement feature extraction. Further, individual blocks are divided into bins comprising sorted intensity values. The feature vector of a particular block is computed by the average of sorted intensity values of the corresponding bin. The process is repeated for all three color components and the generated feature vectors are combined to form the feature vector of the image.

3.7 Code Example (MATLAB)

#Read an image
```
i=imread('path\imagefile');
```

#Separate the color components
```
r=i(:,:,1); #red component
[m n p]=size(r); #dimension of red component
g=i(:,:,2); #green component
[x y z]=size(g); #dimension of green component
b=i(:,:,3); #blue component
[j k l]=size(b); #dimension of blue component
```

#Intensity values as single dimensional array
```
r1=r(:);
g1=g(:);
b1=b(:);
```

#Sorting of Intensity values
```
r2=sort(r1);
g2=sort(g1);
b2=sort(b1);
```

#Dimension of array
```
[m n p]=size(r2);
```

```
[m1 n1 p1]=size(g2);
[m2 n2 p2]=size(b2);
```

#Computing feature vector for 4 bins

```
for a1 = 0:3
Rav(a1+1)=mean2(r2((a1*m/4+1):(a1*m/4+m/4)));
Gav(a1+1)=mean2(g2((a1*m/4+1):(a1*m/4+m/4)));
Bav(a1+1)=mean2(b2((a1*m/4+1):(a1*m/4+m/4)));
end
```

3.8 Coding Exercise

The previous code example is for designing 4 bins for feature extraction. Refer to the example to complete the following assignment:

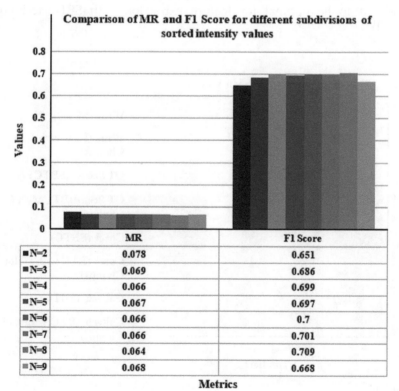

FIGURE 3.4
Comparison of MR and F1-Score for Different Values of N.
[Observation: $N = 8$ has the highest F1-Score and lowest MR.]

- Compute the feature vector using 5, 6, 7, 8 and 9 bins
  Initially, the Wang dataset is considered a testbed to carry out a classification task using KNN Classifier. The classification process is carried out as follows:
- Result comparison for classification is carried out considering the number of bins as $N = 3$ and $N = 2$.
- The increased F1-Score and the reduced MR value is recorded for $N = 3$ compared to $N = 2$.
- Therefore, further subdivisions are initiated to extract feature vectors and to compare the results to the immediate previous subdivision.
- Improved results are recorded until $N = 8$ as reported in Fig. 3.6.
- F1-Score is *0.668* and MR is *0.068* respectively for $N = 9$, whereas the F1-Score is *0.706* and MR is *0.064* for $N = 8$.

Thus, further subdivisions show clear signs of reduced performance. Further subdivisions are therefore found to be unnecessary. $N = 8$ and it is considered to be optimal for feature extraction with SBTC technique (Fig. 3.4).

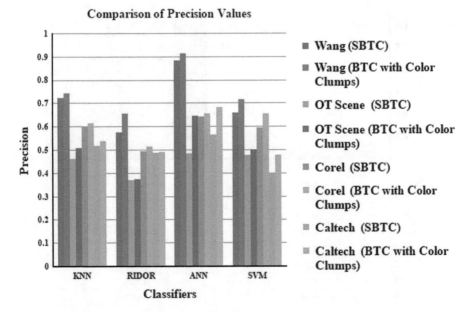

FIGURE 3.5
Comparison of Precision Values of Classification with SBTC and BTC with Color Clumps Method of Feature Extraction for All Four Datasets Using Four Different Classifiers. [Observation: BTC with color clumps has shown better results compared to SBTC.]

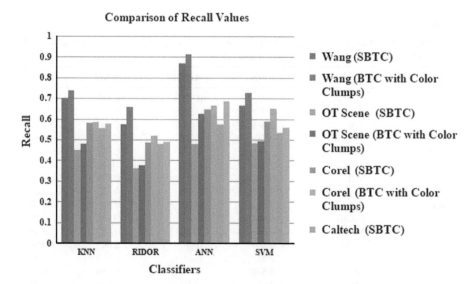

FIGURE 3.6
Comparison of Recall Values of Classification with SBTC and BTC with Color Clumps Method
of Feature Extraction for All Four Datasets Using Four Different Classifiers.
[Observation: BTC with color clumps has shown better results compared to SBTC.]

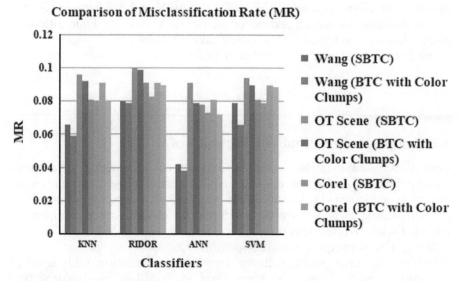

FIGURE 3.7
Comparison of MR with SBTC and BTC with Color Clumps Method of Feature Extraction for
All Four Datasets Using Four Different Classifiers.
[Observation: BTC with color clumps has shown less MR compared to SBTC.]

TABLE 3.3

Classification Performance by Feature Extraction Using SBTC with Four Different Public Datasets Using Four Different Classifiers

Datasets	Metrics	KNN	RIDOR	ANN	SVM
Wang	Precision	0.72	0.575	0.882	0.657
	Recall	0.7	0.572	0.869	0.664
	MR	0.066	0.08	0.042	0.079
	F1-Score	0.7	0.573	0.869	0.657
OT-Scene	Precision	0.508	0.374	0.644	0.476
	Recall	0.479	0.375	0.625	0.481
	MR	0.092	0.1	0.079	0.094
	F1-Score	0.475	0.375	0.625	0.467
Corel	Precision	0.613	0.492	0.64	0.595
	Recall	0.584	0.487	0.645	0.586
	MR	0.081	0.091	0.073	0.081
	F1-Score	0.569	0.489	0.64	0.57
Caltech	Precision	0.537	0.487	0.682	0.4
	Recall	0.578	0.477	0.686	0.532
	MR	0.08	0.091	0.072	0.09
	F1-Score	0.537	0.482	0.686	0.443

[**Observation:** All four datasets have highest classification results with ANN Classifier.]

Hereafter, calculation of average values for diverse metrics is carried out to observe Precision, Recall, MR and F1-Score for classification with four different classifiers—namely, KNN, RIDOR, ANN and SVM using 10-fold cross-validation for feature extraction using SBTC ($N = 8$). Four different public datasets are tested for the classification results.

The comparative results are given in Table 3.3.

3.9 Comparison of Proposed Techniques

Color features are extracted using two different color averaging techniques, namely BTC with color clumps and SBTC, to enhance the performance of content-based image classification (Figs. 3.5 and 3.6). A comparative study of the two techniques is performed considering the classification performances recorded with diverse metrics—namely, Precision, Recall, MR and F1-Score.

The graphical comparisons are shown in Figs. 3.7–3.10.

The results have established the fact that classification with feature extraction using BTC with color clumps has outclassed the results for SBTC.

BTC with color clumps has logically separated the blocks at homogeneous intensity intervals, e.g., 0–127 and 128–255 for two clumps to calculate the feature vectors. The color clumps have carried out formation of disjoint subsets

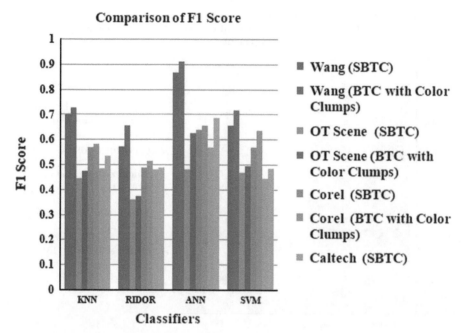

FIGURE 3.8
Comparison of F1-Score with SBTC and BTC with Color Clumps Method of Feature Extraction for All Four Datasets Using Four Different Classifiers.
[Observation: BTC with color clumps has shown better results compared to SBTC.]

of gray values with their distribution approximation. Thus, the disjoint subsets do not exhibit any common values, which in turn have assisted in computing features from large varieties of gray values with higher detailing about the test image. This has ultimately boosted extraction of features by meaningful embracing of significant details of the image compared to SBTC, which has taken a straight-forward approach in direct sorting of gray values without separating it into bins for feature extraction. Thus, there was no creation of disjoint subsets for the bins in SBTC. This has also hampered any logical correlation among the sorted gray values for capturing greater image detail.

3.10 Comparison with Existing Techniques

Classification performance of feature extraction using BTC with color clumps is compared to the benchmarked techniques [9–13] because of its superior performance compared to SBTC in Section 3.9. Classification comparison with KNN classifier of BTC with color clumps to that of state-of-the-art techniques is carried out with Wang dataset.

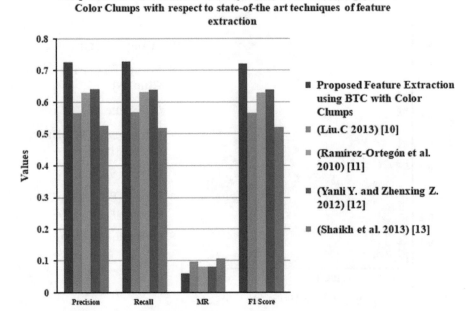

FIGURE 3.9

Comparison of Average Precision, Average Recall, Average Misclassification Rate and Average F1-Score of Existing Techniques wrt BTC with Color Clumps Technique of Feature Extraction for Classification Done with KNN Classifier.

[Observation: Classification by feature extraction using BTC with color clumps has the highest Precision, Recall and F1-Score values and lowest MR value compared to the existing techniques.]

The comparison is shown in Fig. 3.11

The comparison showcased the supremacy of the proposed technique over state of the art.

Finally, a comparison of feature extraction time for the techniques compared in Fig. 3.11 is given in Fig. 3.10. The results reveal least feature extraction time consumed by the proposed technique.

3.11 Statistical Significance

The chapter has demonstrated a test to establish statistical significance of the enhanced classification results recorded with the proposed technique. The test has computed the *p*-values of the existing techniques with respect

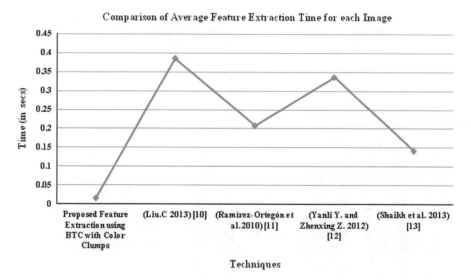

FIGURE 3.10
Comparison of Average Time Consumed for Feature Extraction.
[Observation: Feature extraction with BTC with color clumps has consumed minimum average time per image among all the proposed techniques.]

to the proposed technique. The *t*-test has revealed the actual difference between two means corresponding to the variation in the data of the proposed technique and the existing techniques. The test is carried out by designing a null hypothesis to authenticate the statistical significance of performance enhancement based on Precision values.

HYPOTHESIS 3.1: *There is no difference between the Precision values of classification result obtained by feature extraction using BTC with color clumps with respect to the existing techniques*

The values of *mean, standard deviation (SD)* and *standard error of the mean (SEM)* of the Precision values of classification of different categories for feature extraction using BTC with color clumps is given in Table 3.4.

Table 3.5 shows the computed values of *t* as *t*-calc and the *p*-values. The *p*-values generated after the comparison of Precision values for classification by feature extraction with the existing techniques with respect to the proposed technique have revealed significant difference in Precision results for feature extraction with all the existing techniques. Therefore, the null hypothesis of equal Precision rates for the above mentioned existing algorithm compared to the proposed algorithm is rejected for all the existing techniques given in Table 3.5. Further, the *mean, standard deviation* and *standard error of the mean (SEM)* for precision values for classification

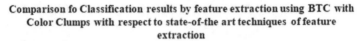

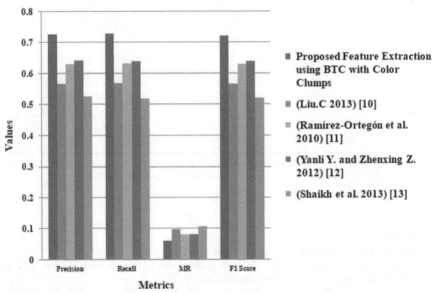

FIGURE 3.11

Comparison of Average Precision, Average Recall, Average misclassification Rate and Average F1-Score of Existing Techniques wrt BTC with Color Clumps Technique of Feature Extraction for Classification Done with KNN Classifier.

[Observation: Classification by feature extraction using BTC with color clumps has the highest Precision, Recall, F1-Score values and lowest MR value compared to the existing techniques.]

by BTC with color clumps technique of feature extraction given in Table 3.4 is compared to that of the existing techniques in Table 3.5. It is observed that the mean Precision value for the novel technique has exceeded the *mean* Precision values of existing techniques. The value of *SD* and *SEM* for the novel technique is also less compared to the existing techniques. Therefore, it can be concluded that the introduced technique of

TABLE 3.4

Value of Mean Standard Deviation and Standard Error of the Mean for Precision Values of Classification by Feature Extraction Using BTC with Color Clumps

Proposed Technique	Mean	Standard Deviation (SD)	Standard Error of the Mean (SEM)
Feature Extraction using BTC with Color Clumps	0.747	0.154	0.051

TABLE 3.5

t-test for Evaluating the Significance of Feature Extraction Using BTC with Color Clumps

Comparison	Mean	Standard Deviation (SD)	Standard Error of the Mean (SEM)	*t-calc*	*p*-Value	Significance
Liu, 2013	0.626	0.22	0.073	2.85	0.02	*Significant*
Ramírez-Ortegón et al., 2010	0.601	0.24	0.08	2.47	0.03	*Significant*
Yanli and Zhenxing, 2012	0.638	0.223	0.074	2.95	0.02	*Significant*
Shaikh et al. 2013	0.595	0.214	0.071	4.28	0.002	*Significant*

[**Observation:** *p* value have indicated significant difference in Precision results for classification with all the existing feature extraction techniques compared to the feature extraction using BTC with color clumps.]

feature extraction has greater consistency compared to the existing techniques. Hence, the proposed method has shown statistically significant contribution for enhanced classification compared to the existing techniques.

The *t-test* is performed with four different comparisons. Therefore, a *post hoc* analysis is felt necessary to be conducted to accept the significance of the derived results using the *Bonferroni correction*. The value of α is considered to be *0.05* for the *t-test*. The expression to calculate the *Bonferroni correction* value is given below:

$$Bonferroni\_\alpha = \frac{\alpha}{Number\_of\_tests}$$

where $\alpha = 0.05$ *Number_of_tests* = 4 Therefore, *Bonferroni_α* = 0.013

Now, the *p- values* in Table 3.5 are compared to the newly calculated *Bonferroni_α* value of 0.013. The comparison shows that none of the *p-values* in Table 3.5 is smaller than *Bonferroni_α* value, although they have shown significant statistical difference during *t-test*.

Chapter Summary

Two different color averaging-based feature extraction techniques—namely, feature extraction using block truncation coding (BTC) with color clumps and feature extraction with sorted block truncation coding (SBTC)—have been introduced in this chapter. The techniques are tested for classification performance with popular public datasets—namely, Wang dataset, OT-Scene dataset, Caltech dataset, etc. Henceforth, classification results for feature extraction using BTC with color clumps is compared with feature extraction using SBTC. The former technique has outclassed the latter. Hence, the classification results by feature extraction using BTC with color clumps technique is compared to the existing techniques. The results have revealed statistically significant improved classification with the proposed technique of feature extraction compared to the existing techniques.

References

1. Panda, D. K. and Meher, S., 2015. Detection of moving objects using fuzzy color difference histogram based background subtraction. *IEEE Signal Processing Letters*, 23(1): 45–49.
2. Bannai, N., Agathos, A., and Fisher, R. B., 2004. Fusing multiple color images for texturing models. In Proceedings. 2nd International Symposium on 3D Data Processing, Visualization and Transmission, 2004. 3DPVT 2004, *IEEE*, pp. 558–565.
3. Soni, D. and Mathai, K. J., 2015, April. An efficient content based image retrieval system based on color space approach using color histogram and color correlogram. In 2015 Fifth International Conference on Communication Systems and Network Technologies, *IEEE*, pp. 488–492.
4. Lu, T.-C., Lu, T.-C., Chang, C.-C. and Chang, C.-C., 2007. Color image retrieval technique based on color features and image bitmap. *Information Processing & Management*, 43(2): 461–472.
5. Nazir, A., Ashraf, R., Hamdani, T. and Ali, N., 2018, March. Content based image retrieval system by using HSV color histogram, discrete wavelet transform and edge histogram descriptor. In 2018 International Conference on Computing, Mathematics and Engineering Technologies, iCoMET, *IEEE*, pp. 1–6.
6. Li, J., Tian, Y. and Cassidy, M. J., 2014. Failure mechanism and bearing capacity of footings buried at various depths in spatially random soil. *Journal of Geotechnical and Geoenvironmental Engineering*, 141(2): 04014099-1-04014099-11.
7. Walia, E., Vesal, S. and Pal, A., 2014. An effective and fast hybrid framework for color image retrieval. Sensing and Imaging, *Springer*, 15(1): 1–23.
8. Delp, E. and Mitchell, O., 1979. Image compression using block truncation coding. *IEEE transactions on Communications*, 27(9): 1335–1342.
9. Han, J. and Kamber, M. 2001. Data mining: Concepts and techniques. *The Morgan Kaufmann Series in Data Management Systems*, pp. 89–90.
10. Liu. C., 2013, A new finger vein feature extraction algorithm. *In IEEE 6th International Congress on Image and Signal Processing (CISP)*, pp. 395–399.
11. Ramírez-Ortegón, M. A. and Rojas, R., 2010. Unsupervised evaluation methods based on local gray-intensity variances for binarization of historical

documents. *In Proceedings of the International Conference on Pattern Recognition,* pp. 2029–2032.

12. Yanli Y. and Zhenxing Z., 2012. A novel local threshold binarization method for QR image. *In IET International Conference on Automatic Control and Artificial Intelligence (ACAI),* pp. 224–227.

13. Shaikh, S. H., Maiti, A. K. and Chaki, N., 2013. A new image binarization method using iterative partitioning. *Machine Vision and Applications,* 24(2): 337–350.

4

Content-Based Feature Extraction: Image Binarization

4.1 Prelude

Binarization is a popular technique of converting the intensity values in images to 0 and 1. However, it is carried out by selecting a threshold for conversion of the gray values. Binarization is visualized as a robust feature extraction technique and has exhibited high accuracy for content-based image classification. The region of interest can be differentiated efficiently from its background by means of the adept platform provided in image binarization. However, selection of a threshold value has a crucial role to determine the appropriateness of the binarization technique. Therefore, special care must be realized while computing threshold value since it is instrumental in differentiating the region of interest from its background. Threshold selection is broadly classified into three different categories:

- Mean threshold selection
- Local threshold selection
- Global threshold selection

This chapter contains an elaborate illustration for innovatively using mean threshold selection and local threshold selection to extract meaningful features from images based on block truncation coding described in Chapter 3. Working with global threshold selection is left as an exercise for the reader. The novel approaches of feature extraction discussed in this chapter extract lightweight features independent of the size of the image input. This flexibility addresses the computational complexity and space complexity favorably and reduces the computational overhead of the entire system. An illustration of a binarized image is provided in Fig. 4.1.

(a)   (b)

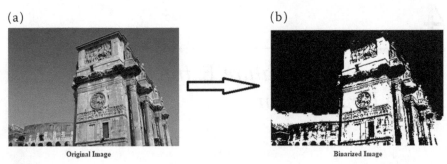

FIGURE 4.1
Image Binarization.

4.2 Feature Extraction Using Mean Threshold Selection

This section introduces mean threshold-based binarization for feature extraction. It introduces four different varieties to demonstrate the process with the help of code examples. Stepwise methodology for computing multilevel mean threshold features is given in Section 4.2.1.

4.2.1 Feature Extraction with Multilevel Mean Threshold Selection

The term "multilevel" refers to more than one stage. Hence, it is quite evident that the feature extraction is carried out with multiple stages of mean threshold calculation. The dataset considered as a testbed in this case has required four different stages of mean threshold selection to design robust descriptors with the technique [1]. The stages are named BTC Level 1, BTC Level 2, BTC Level 3 and BTC Level 4 as shown in Fig. 4.2.

The steps for feature extraction follow:

- Input an image, I, having the three different color components R, G and B of size $m * n$ each.
- The mean threshold value T_x is to be calculated for each pixel in each color component R, G and B.
- The binary image map for each pixel of the given image is to be computed.

$$Bitmapx(i, j) = \begin{cases} 1.. \ if..x(i, j) > Tx \\ 0. \ if. \ x(i, j) < Tx \end{cases}$$

/*x = R, G and B */

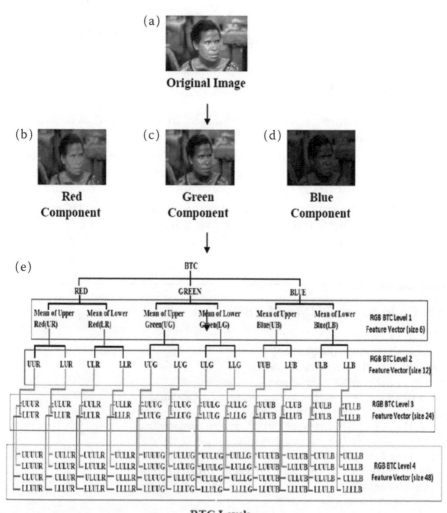

FIGURE 4.2
Stages of Multilevel Block Truncation Coding.

- Two clusters are formed with gray values that are represented with *1* and *0* respectively after comparing them to the threshold.
- Calculate the mean of each cluster formed in the previous step to generate image features for the given image.

$$Xupmean = \frac{1}{\sum_{i=1}^{m}\sum_{j=1}^{n} BitMap_x(i, j)} * \sum_{i=1}^{m}\sum_{j=1}^{n} BitMap_x(i, j) * X(i, j)$$

$$Xlomean = \frac{1}{m * n - \sum_{i=1}^{m}\sum_{j=1}^{n} Bitmap_x(i,j)} * \sum_{i=1}^{m}\sum_{j=1}^{n} (1 - BitMap_x(i,j)) * X(i,j)$$

/*x = R, G and B */

- The *six* feature vectors thus obtained in *level 1 BTC* acts as six thresholds for *level 2*.
- Repeat steps 4 and 5 to compute *Xupmean* and *Xlomean* for each of six thresholds, which will calculate *twelve* elements of feature vectors in *level 2 BTC*.
- Repeat steps 4 and 5 to compute *twenty-four* and *forty-eight* elements of feature vector respectively for *level 3 BTC* and *level 4 BTC*.

4.3 Code Example (MATLAB®)

#Read an image
```
i=imread('path\imagefile');
  #Separate the color components
r=i(:,:,1); #red component
[m n p]=size(r); #dimension of red component
g=i(:,:,2); #green component
[x y z]=size(g); #dimension of green component
b=i(:,:,3); #blue component
[j k l]=size(b); #dimension of blue component
  #Calculate mean threshold for each color component
r_thrsh=mean2(r);
g_thrsh =mean2(g);
b_thrsh =mean2(b);
  #Calculate mean of intensity values higher than threshold
for a1=1:m
for a2=1:n
if r(a1,a2)>= r_thrsh #Comparing intensity values to threshold
c1=c1+1;
s1=s1+r(a1,a2); #Adding up intensity values higher than threshold
end
end
end
if c1~=0
if s1~=0
supavred=s1/c1; #Calculating mean of intensity values higher than
threshold
```

```
end
end
if c1==0
if s1==0
supavred=0;
end
end
for a3=1:m
for a4=1:n
if r(a3,a4)< r_thrsh #Comparing intensity values to threshold
c2=c2+1;
s2=s2+i1(a3,a4); #Adding up intensity values higher than threshold
end
end
end
if c2~=0
if s2~=0
sloavred=s2/c2; #Calculating mean of intensity values higher than
threshold
else
sloavred=0;
end
end
```

4.4 Coding Exercise

The previous code example is for designing feature vectors for BTC Level 1 from the Red color component. Refer to the example to complete the following assignments:

- Compute feature vectors for color components Green and Blue.
- Compute feature vectors for BTC Levels 2, 3 and 4.

Comparison of the derived feature vectors at different levels is carried out based on the metrics misclassification rate (MR) and F1-Score of classification with the Wang dataset as in Fig. 4.3.

The comparison in Fig. 4.3 has clearly established BTC Level 3 as the most suitable option for feature extraction using mean threshold selection. This is because the lowest MR and highest F1-Score for classification is observed with the referred level of BTC.

Hereafter, calculation of average values for diverse metrics is carried out to observe Precision, Recall, MR and F1-Score for classification with four different classifiers—namely, K Nearest Neighbor (KNN), Ripple-Down

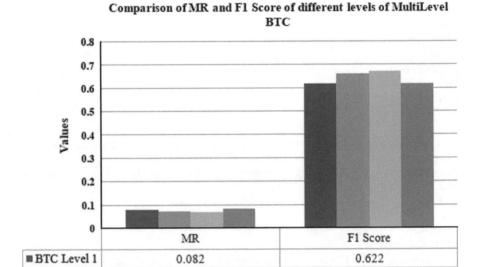

FIGURE 4.3
Comparison of MR and F1-Score for BTC Levels 1, 2, 3 and 4.
[**Observation:** BTC Level 3 has the lowest MR and highest F1-Score.]

Rule (RIDOR), Artificial Neural Network (ANN) and Support Vector Machine (SVM)—for BTC level 3 used for feature extraction on four different public datasets with mean threshold selection. The comparison of classification performances is shown in Table 4.1 (Table 4.1).

4.5 Feature Extraction from Significant Bit Planes Using Mean Threshold Selection

The gray values in an image are represented by decimal numbers. The decimal numbers can be easily represented by binary bits. The binary representation of individual bits has been regarded as bit planes in this section [2].

One can enumerate up to eight different bit planes in each of the test images initiating from the least significant bit to the most significant bit. This is inline to the demonstration given in Fig. 4.4.

TABLE 4.1

Classification Performances by Feature Extraction using Mean Threshold Selection for BTC Level 3 with Four Different Public Datasets using Four Different Classifiers

Datasets	Metrics	KNN	RIDOR	ANN	SVM
Wang	Precision	0.686	0.656	0.796	0.754
	Recall	0.681	0.659	0.8	0.751
	MR	0.07	0.074	0.05	0.054
	F1-Score	0.673	0.656	0.797	0.749
OT-Scene	Precision	0.405	0.33	0.434	0.327
	Recall	0.399	0.329	0.432	0.342
	MR	0.1	0.14	0.095	0.13
	F1-Score	0.395	0.328	0.431	0.324
Corel	Precision	0.489	0.412	0.52	0.427
	Recall	0.5	0.414	0.534	0.44
	MR	0.093	0.097	0.079	0.11
	F1-Score	0.477	0.412	0.52	0.392
Caltech	Precision	0.564	0.624	0.455	0.303
	Recall	0.624	0.627	0.577	0.488
	MR	0.079	0.081	0.091	0.12
	F1-Score	0.543	0.625	0.49	0.351

[Observation: Wang and OT-Scene datasets have the highest classification rate with ANN Classifier; Corel and Caltech datasets have the highest classification rate with KNN Classifier.]

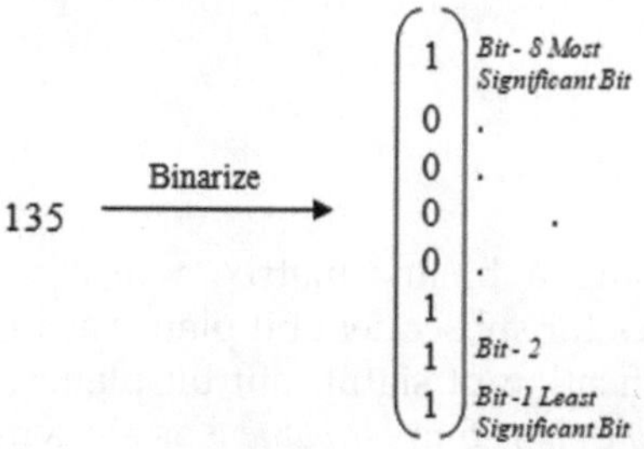

FIGURE 4.4 Binary Representation of Gray Value as Bit Plane.

The expression for bit planes has been given as follows in equation 4.1.

$$I_{bitplane}(p, q) = RM\left\{\frac{1}{2}floor\left[\frac{1}{2^p}I(p, q)\right]\right\} \qquad (4.1)$$

- $I(p,q)$: Original Image
- *RM*: Remainder
- *floor(I)*: Rounding the elements to *I* nearest integers less than or equal to *I*.

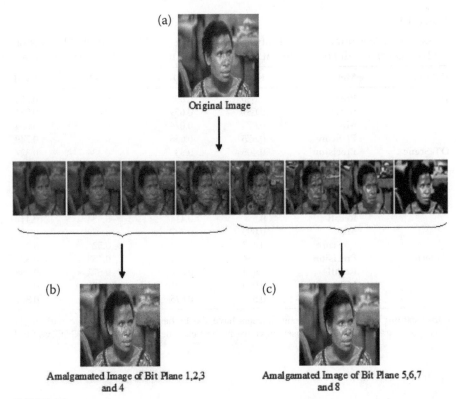

FIGURE 4.5
Steps of Bit Plane Slicing.

Individual bit planes are represented using a binary matrix, which is thereafter instrumental to create image slices for subsequent bit planes as in Fig. 4.5. The process is followed by identification of significant bit planes, which resulted in bit planes ranging from *bit plane 5* to *bit plane 8* as shown in equation 4.2. The selected bit planes are considered for extraction of feature extraction by selecting mean threshold for binarization. The rest of the bit planes are considered to be noisy, shown in Fig. 4.6, and are ignored.

$$bp = bp_{hi} \dots iff \dots bp5 = 1.. \ or..bp6 = 1.. \ or..bp7 = 1.. \ or..bp8 = 1 \quad (4.2)$$

Feature extraction with bit plane slicing are explained in the following steps:

- Follow steps 1 to 5 given in Section 4.2.1 to generate features from the original image.

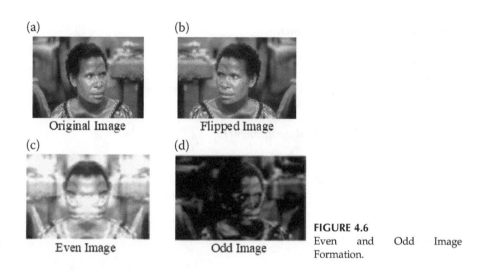

FIGURE 4.6
Even and Odd Image Formation.

- Identify the higher bit planes for each color component starting from *bitplane 5* to *bitplane 8,* which is equal to *1.*

  $$xbp = xbp(bp5 = 1.. \ or..bp6 = 1.. \ or..bp7 = 1.. \ or..bp8 = 1)$$

- /*x = R, G and B */
- Compute mean threshold for the identified significant bit planes for each color component.
- Compute binary image maps for each pixel of significant bit planes for R, G and B respectively.
- Compute the feature vectors $xbitplane_{hi}$ and $xbitplane_{lo}$ for bit planes for all of the three color components using the binary image maps.
- The feature vectors *Xupmean, Xlomean, Xbpupmean* and *Xbplomean* for each color component are associated to that of the original image to form twelve feature vectors altogether for each image in the dataset.

4.6 Code Example (MATLAB)

#Read an image
```
i=imread('path\imagefile');
```

#Separate the color components
```
r=i(:,:,1); #red component
[m n p]=size(r); #dimension of red component
g=i(:,:,2); #green component
[x y z]=size(g); #dimension of green component
b=i(:,:,3); #blue component
[j k l]=size(b); #dimension of blue component
```
#Extracting bit planes (red component)
```
for a1=1:m
for a2=1:n
   #extraction of bit planes
if   bitget(uint8(r(a1,a2)),5)==1   ||bitget(uint8(r(a1,a2)),6)
==1 ||
bitget(uint8(r(a1,a2)),7)==1 || bitget(uint8(r(a1,a2)),8)==1)
rt(a1,a2,:)=r(a1,a2);
end
end
end
```

4.7 Coding Exercise

The previous code example is for designing feature vectors with bit plane slicing from the Red color component. Refer to the example to complete the following assignments:

- *Extract bit planes from Green and Blue color components.*
- *Extract feature vectors from each of the color components using mean threshold selection.*

Hereafter, calculation of average values for diverse metrics is carried out to observe Precision, Recall, MR and F1-Score for classification with four different classifiers—namely, KNN, RIDOR, ANN and SVM—from four different public datasets (Table 4.2).

4.8 Feature Extraction from Even and Odd Image Varieties Using Mean Threshold Selection

This section discusses another example of mean threshold selection for enhancing the robustness of extracted feature vectors with augmented data. The flipped versions of the images are used for data augmentation, which is

TABLE 4.2

Classification Performances by Feature Extraction using Mean Threshold Selection from Bit Planes with Four Different Public Datasets using Four Different Classifiers

Datasets	Metrics	KNN	RIDOR	ANN	SVM
Wang	Precision	0.729	0.654	0.832	0.77
	Recall	0.729	0.654	0.831	0.77
	MR	0.06	0.072	0.042	0.05
	F1-Score	0.722	0.653	0.831	0.765
OT-Scene	Precision	0.456	0.355	0.51	0.449
	Recall	0.426	0.359	0.508	0.436
	MR	0.096	0.012	0.089	0.094
	F1-Score	0.424	0.357	0.509	0.422
Corel	Precision	0.578	0.493	0.64	0.575
	Recall	0.539	0.49	0.648	0.558
	MR	0.09	0.091	0.079	0.082
	F1-Score	0.508	0.491	0.638	0.534
Caltech	Precision	0.411	0.435	0.514	0.371
	Recall	0.503	0.42	0.572	0.505
	MR	0.097	0.098	0.078	0.1
	F1-Score	0.412	0.427	0.533	0.393

[**Observation:** All four datasets have maximum classification rate with ANN Classifier.]

envisioned to enhance the richness of the test image using the same available data.

RGB color space is considered to carry out the technique and the principle of Block truncation Coding (BTC) [3] is followed to initiate feature extraction from the three color components—namely, Red (R), Green (G) and Blue (B). Two different varieties of images are formed—namely. even image and odd image—by following the equations 4.3 and 4.4 respectively.

$$Even\ Image = \frac{(Original\ Image + Flipped\ Image)}{2} \tag{4.3}$$

$$Odd\ Image = \frac{(Original\ Image - Flipped\ Image)}{2} \tag{4.4}$$

Primarily, flipping of the test images is performed across X- and Y-axes as in Fig. 4.6. This creates the a new flipped version of the original image and is added to the original image to create an even image. Subtracting the flipped version from the original image creates the odd image. Both the actions are shown in Fig. 4.6.

Once the preprocessing of the test images is over, the rest of the process is similar to the steps of feature extraction using mean threshold selection discussed in Section 4.2.1.

Comparison of MR and F1 Score for Original, Even and Odd Image varieties

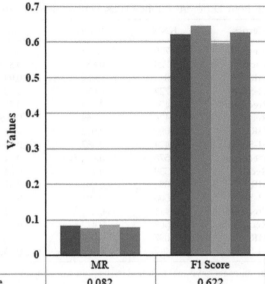

	MR	F1 Score
▪ Original Image	0.082	0.622
▪ Original Image + Even Image	0.076	0.646
▪ Original Image + Odd Image	0.084	0.598
▪ Original + Even Image + Odd Image	0.078	0.627

Metrics

FIGURE 4.7
Comparison of MR and F1-Score for Original, Even and Odd Image Varieties. **[Observation:** Original + Even Image has the lowest MR and highest F1-Score.]

Primarily, each variety of test images created after preprocessing is compared against each other for classification performance using the metrics MR and F1-Score in Fig. 4.7. The Wang dataset is used for the purpose. Observations in Fig. 4.7 showcase feature extraction with Original + Even Image variety having the minimum MR and maximum F1-Score compared to the rest of the techniques.

4.9 Code Example (MATLAB)

#Read an image
```
i=imread('path\imagefile');
```
  #Separate the color components

```
r=i(:,:,1); #red component
[m n p]=size(i1); #dimension of red component
g=i(:,:,2); #green component
[x y z]=size(i3); #dimension of green component
b=i(:,:,3); #blue component
[j k l]=size(i6); #dimension of blue component
  #Flipping the color components
flp_r =fliplr(r);
flp_g =fliplr(g);
flp_b =fliplr(b);
  #Creating Even and Odd varieties of each component
r_evn = (r + flp_r)/2;
g_evn = (g + flp_g)/2;
b_evn = (b + flp_b)/2;
r_odd = (r - flp_r)/2;
g_odd = (g - flp_g)/2;
b_odd = (b - flp_b)/2;
```

4.10 Coding Exercise

The previous code example is for generating odd and even varieties of test images. Refer to the example to complete the following assignments:

- *Extract features from odd and even varieties of images using mean threshold selection.*
- *Test classification results for each variety.*

Hereafter, calculation of average values for diverse metrics is carried out to observe Precision, Recall, MR and F1-Score for classification with four different classifiers—namely, KNN, RIDOR, ANN, and SVM—from four different public datasets using features extracted with mean threshold selection from Original + Even Image variety (Table 4.3).

4.11 Feature Extraction with Static and Dynamic Ternary Image Maps Using Mean Threshold Selection

This section deals with extraction of feature vectors with ternary image maps following the logic of mean threshold selection [4]. The techniques are known as Static Thepade's Ternary BTC (STTBTC) and Dynamic Thepade's

TABLE 4.3

Classification Performances by Feature Extraction using Mean Threshold Selection from Even Image Variety with Four Different Public Datasets using Four Different Classifiers

Datasets	Metrics	KNN	RIDOR	ANN	SVM
Wang	Precision	0.647	0.585	0.694	0.666
	Recall	0.649	0.589	0.699	0.678
	MR	0.076	0.078	0.066	0.071
	F1-Score	0.646	0.587	0.695	0.667
OT-Scene	Precision	0.391	0.314	0.394	0.283
	Recall	0.392	0.309	0.397	0.317
	MR	0.099	0.15	0.098	0.18
	F1-Score	0.387	0.31	0.39	0.272
Corel	Precision	0.495	0.417	0.515	0.437
	Recall	0.48	0.417	0.527	0.442
	MR	0.093	0.096	0.089	0.097
	F1-Score	0.453	0.417	0.513	0.403
Caltech	Precision	0.437	0.483	0.546	0.401
	Recall	0.522	0.51	0.57	0.502
	MR	0.094	0.091	0.076	0.098
	F1-Score	0.446	0.493	0.551	0.394

[**Observation:** All four datasets have maximum classification rate with ANN Classifier.]

Ternary BTC (DTTBTC). Both methods have followed the BTC-based technique of feature extraction by separating the Red (R), Green (G) and Blue (B) color components from each of the test images. The following subsections A and B discuss both techniques.

The process of feature extraction with STTBTC are explained in the following steps:

- Follow steps 1 and 2 given in Section 4.2.1 to generate threshold T_x for each color component R, G and B.

$$Tx = (1/m * n) * \sum_{i=1}^{m} \sum_{j=1}^{n} x(i, j)$$

- /* x = R, G and B */

- One overall luminance threshold T is to be generated by averaging T_x.

$$T = \sum \frac{Tx}{3}$$

/* x=R, G and B */

- Individual color threshold for each color component is to be calculated. $T_x lo$ is considered to be the lower threshold, and $T_x hi$ is regarded as the higher threshold by altering the value of n from 1 to 5.

$$T_x lo = T_x - n \,|\, T_x - T_{overall} \,| \,, \; T_x hi = T_x + n \,|\, T_x - T_{overall} \,|$$

/* x=R, G and B */
- Compute ternary image maps.

$$T_{mx}(i, j) = \begin{cases} 1 & if & x(i, j) > T_x hi \\ 0 & if & T_x lo < =x(i, j) < =T_x hi \\ -1 & if & x(i, j) < T_x lo \end{cases}$$

/* x=R, G and B */
- Generate image features tUx, tMx and tLx for ternary BTC.

$$tUx = \frac{1}{\sum_{i=1}^{m}\sum_{j=1}^{n} Tmx(i, j)\, iff Tmx(i, j) = 1} * \sum_{i=1}^{m}\sum_{j=1}^{n} x(i, j),\, iff Tmx(i, j) = 1$$

$$tMx = \frac{1}{\sum_{i=1}^{m}\sum_{j=1}^{n} Tmx(i, j)\, iff Tmx(i, j) = 0} * \sum_{i=1}^{m}\sum_{j=1}^{n} x(i, j),\, iff Tmx(i, j) = 0$$

$$tLx = \frac{1}{\sum_{i=1}^{m}\sum_{j=1}^{n} Tmx(i, j)\, iff Tmx(i, j) = -1} * \sum_{i=1}^{m}\sum_{j=1}^{n} x(i, j),\, iff Tmx(i, j) = -1$$

/* x=R, G and B */

The other technique, named DTTBTC, has computed the degree values of n in a different way compared to the Static Thepade's Ternary BTC technique. Calculation of the absolute ratio of the threshold for each color component to the overall threshold value of luminance is performed to determine the value of n as given in equation 4.5. The computation has resulted in three different values for three color components according to equation 4.12. The values are compared to each other for choosing the largest value out of the three [4]. The largest value is further considered as the degree value n to be associated dynamically for threshold calculation.

$$
n = \begin{cases} \left| \dfrac{T_r}{T_{overall}} \right|, & iff \, |T_g| < |T_r| > |T_b| \\[2em] \left| \dfrac{T_g}{T_{overall}} \right|, & iff \, |T_r| < |T_g| > |T_b| \\[2em] \left| \dfrac{T_b}{T_{overall}} \right|, & iff \, |T_r| < |T_b| > |T_g| \end{cases} \tag{4.5}
$$

Comparison of the two feature extraction processes considering the Wang dataset as a testbed is given in Fig. 4.8 based on MR and F1-Score.

The comparison shown in Fig. 4.8 shows that DTTBTC has outclassed STTBTC with minimized MR and maximized F1-Score.

4.12 Code Example (MATLAB)

```
#Read an image
i=imread('path\imagefile');
  #Separate the color components
r=i(:,:,1); #red component
[m n p]=size(r); #dimension of red component
g=i(:,:,2); #green component
[x y z]=size(g); #dimension of green component
b=i(:,:,3); #blue component
[j k l]=size(b); #dimension of blue component
  #calculation of mean threshold for STTBTC
Tavred=mean2(r);
Tavgrn=mean2(g);
Tavblu=mean2(b);
  #calculation of threshold for STTBTC
#calculating overall luminance threshold
Tav= (Tavred+Tavgrn+Tavblu)/3;
#calculating lower and higher threshold values for each color
component
Tavrl= Tavred- n*(abs(Tavred-Tav));
Tavrh= Tavred+ n*(abs(Tavred-Tav));
#n = 1 to 5
Tavgl= Tavgrn- n*(abs(Tavgrn-Tav));
Tavgh= Tavgrn+ n*(abs(Tavgrn-Tav));
Tavbl= Tavblu- n*(abs(Tavblu-Tav));
Tavbh= Tavblu+ n*(abs(Tavblu-Tav));
  #calculation of threshold for DTTBTC
Trd=Tavred/Tav;
```

```
Tgd=Tavgrn/Tav;
Tbd=Tavblu/Tav;
```
  #comparing the values dynamically for computing 'n'
```
if (Tgd<Trd>Tbd)
n=Trd;
elseif(Trd<Tgd>Tbd)
n=Tgd;
elseif(Tgd<Tbd>Trd)
n=Tbd;
end
```
  #determining the higher and lower threshold values with dynamically computed 'n'
```
Tavrl= Tavred-n*(abs(Tavred-Tav));
Tavrh= Tavred+n*(abs(Tavred-Tav));
Tavgl= Tavgrn-n*(abs(Tavgrn-Tav));
Tavgh= Tavgrn+n*(abs(Tavgrn-Tav));
Tavbl= Tavblu-n*(abs(Tavblu-Tav));
Tavbh= Tavblu+n*(abs(Tavblu-Tav));
```

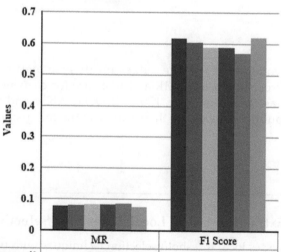

Comparison of MR and F1 Score for STTBTC and DTTBTC

	MR	F1 Score
■ STTBTC (Degree=1)	0.076	0.617
■ STTBTC (Degree=2)	0.079	0.605
■ STTBTC (Degree=3)	0.082	0.589
■ STTBTC (Degree=4)	0.082	0.588
■ STTBTC (Degree=5)	0.086	0.569
■ DTTBTC	0.075	0.621

Metrics

FIGURE 4.8 Comparison of MR and F1-Score of STTBTC to DTTBTC. **[Observation:** DTTBTC has the lowest MR and highest F1-Score.]

TABLE 4.4

Classification Performances by Feature Extraction using Mean Threshold Selection for DTTBTC with Four Different Public Datasets using Four Different Classifiers

Datasets	Metrics	KNN	RIDOR	ANN	SVM
Wang	Precision	0.621	0.538	0.667	0.619
	Recall	0.62	0.541	0.67	0.626
	MR	0.075	0.08	0.072	0.078
	F1-Score	0.621	0.538	0.666	0.616
OT-Scene	Precision	0.421	0.338	0.438	0.379
	Recall	0.418	0.337	0.439	0.372
	MR	0.092	0.097	0.09	0.094
	F1-Score	0.411	0.338	0.437	0.349
Corel	Precision	0.509	0.535	0.451	0.475
	Recall	0.503	0.542	0.45	0.48
	MR	0.09	0.084	0.089	0.087
	F1-Score	0.48	0.531	0.49	0.452
Caltech	Precision	0.444	0.49	0.534	0.432
	Recall	0.535	0.485	0.576	0.517
	MR	0.091	0.089	0.078	0.095
	F1-Score	0.447	0.487	0.543	0.405

[**Observation:** Wang dataset, OT-Scene dataset and Caltech dataset have highest classification rate with ANN Classifier; Corel dataset has highest classification rate with RIDOR Classifier.]

Hereafter, calculation of average values for diverse metrics is carried out to observe Precision, Recall, MR and F1-Score for classification with four different classifiers—namely, KNN, RIDOR, ANN and SVM from four different public datasets with features extracted using DTTBTC (Table 4.4).

4.13 Feature Extraction Using Local Threshold Selection

This section discusses local threshold selection techniques with two popular methods—namely, Niblack's local threshold selection and Sauvola's local threshold selection.

Applications of the local threshold selection techniques are carried out on bit planes of test images and on the original images.

Luminance of images is often found to be uneven, causing major hindrance in selecting a single threshold for the entire image. However, an efficient solution is to segment the image into *nxn* divisions for local threshold selection corresponding to a particular division.

Both the techniques of local threshold selection mentioned in this section have considered both mean and standard deviation for computing the thresholds. This has taken care of the variance of the gray values in the image data.

In the case of Niblack's threshold selection, determination of the threshold is based on the local mean $m(i,j)$ and standard deviation $\sigma(i,j)$. A window size of (25×25) is considered. The threshold is given by $T(i,j) = m(i,j) + k \cdot \sigma(i,j)$. Here, k is a constant having a value between 0 and 1 and is considered to be 0.6 in the proposed method. The value of k and the size of the sliding window determined the quality of binarization.

Sauvola's technique is an improvement over Niblack's method for local threshold selection and is useful for images with faded backgrounds.

Both the techniques are available in the form of *functions()* in the MATLAB library. They can be applied directly to the code example provided in this chapter for feature vector extraction using local threshold selection techniques.

4.14 Code Example (MATLAB)

```
#Read an image
i=imread('path\imagefile');
  #Separate the color components
r=i(:,:,1); #red component
[m n p]=size(r); #dimension of red component
g=i(:,:,2); #green component
[x y z]=size(g); #dimension of green component
b=i(:,:,3); #blue component
[j k l]=size(b); #dimension of blue component
  #Extracting bit planes (red component)
for a1=1:m
for a2=1:n
if bitget(uint8(r(a1,a2)),5)==1 ||bitget(uint8(r(a1,a2)),6)==1
|| bitget(uint8(r(a1,a2)),7)==1 || bitget(uint8(r(a1,a2)),8)==1
rt(a1,a2,:)=r(a1,a2);
end
end
end
BWr = NIBLACK(rt, [25 25], 0.6, 0, 10);
```

4.15 Coding Exercise

The previous code example is for generating local thresholds using Niblack's technique. Refer to the example to complete the following assignments):

- *Extract features from Green and Blue color components using Niblack's technique.*
- *Extract features from Red, Green and Blue color components using Sauvola's technique.*

The demonstrations of application for these techniques are given in Figs. 4.9 and 4.10.

Hereafter, calculation of average values for diverse metrics is carried out to observe Precision, Recall, MR and F1-Score for classification with four different classifiers—namely, KNN, RIDOR, ANN and SVM—from four different public datasets with features extracted using Niblack's and Sauvola's local threshold selection (Tables 4.5 and 4.6).

4.16 Comparing the Discussed Techniques for Performance Evaluation

This chapter discusses six different techniques of feature extraction using binarization from the test images. Four different metrics of comparisons are involved for performance evaluation of the techniques. The metrics are Precision, Recall, MR and F1-Score. The techniques are applied on four different datasets—namely, Wang dataset, OT-Scene dataset, Corel dataset and Caltech dataset. Classification is carried out with four different classifiers—namely, KNN Classifier, RIDOR Classifier, ANN Classifier and SVM Classifier.

The performance comparisons are given in Figs. 4.11–4.14.

The comparative results shown in Figs. 4.11–4.14 have recorded the highest classification performance for feature extraction by binarization with Sauvola's local threshold selection.

4.17 Comparison with Existing Techniques

Classification performance for feature extraction by binarization with Sauvola's local threshold selection is compared to the benchmarked

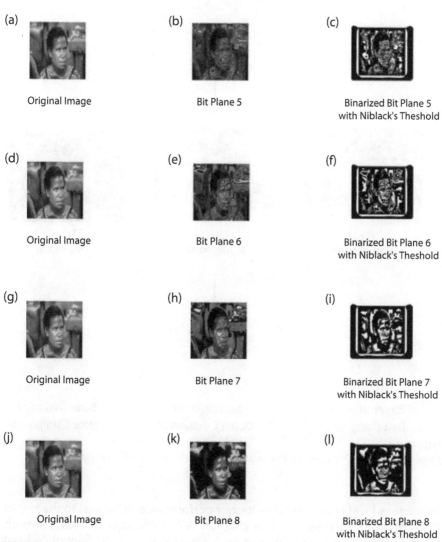

(a)

Original Image

(b)

Bit Plane 5

(c)

Binarized Bit Plane 5
with Niblack's Theshold

(d)

Original Image

(e)

Bit Plane 6

(f)

Binarized Bit Plane 6
with Niblack's Theshold

(g)

Original Image

(h)

Bit Plane 7

(i)

Binarized Bit Plane 7
with Niblack's Theshold

(j)

Original Image

(k)

Bit Plane 8

(l)

Binarized Bit Plane 8
with Niblack's Theshold

FIGURE 4.9
Binarization of Bit Planes using Niblack's Threshold Selection.

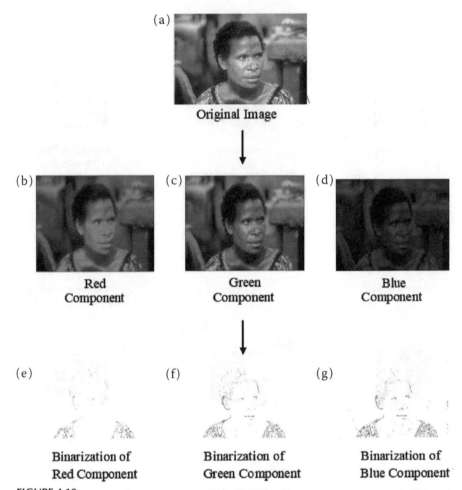

(a)

Original Image

(b)   (c)   (d)

Red   Green   Blue
Component   Component   Component

(e)   (f)   (g)

Binarization of   Binarization of   Binarization of
Red Component   Green Component   Blue Component

FIGURE 4.10
Process of Feature Extraction by Binarization with Sauvola's Local Threshold Selection.

techniques [5–8] because of its superior performance compared to the rest of the techniques discussed in this chapter. A classification comparison with KNN Classifier of feature extraction by binarization with Sauvola's local threshold selection to that of state-of-the-art techniques is carried out with Wang dataset. The comparison is shown in Fig. 4.15.

The comparison showcased the supremacy of the proposed technique over state of the art.

Finally, a comparison of feature extraction time for the techniques is given in Fig. 4.16. The results reveal least feature extraction time consumed by the proposed technique.

TABLE 4.5

Classification Performances by Feature Extraction using Niblack's Local Threshold Selection from Significant Bit Planes with Four Different Public Datasets using Four Different Classifiers

Datasets	Metrics	KNN	RIDOR	ANN	SVM
Wang	Precision	0.676	0.599	0.738	0.615
	Recall	0.68	0.601	0.742	0.634
	MR	0.072	0.079	0.057	0.076
	F1-Score	0.67	0.599	0.738	0.619
OT-Scene	Precision	0.443	0.354	0.456	0.378
	Recall	0.427	0.352	0.449	0.401
	MR	0.091	0.096	0.094	0.093
	F1-Score	0.423	0.353	0.448	0.377
Corel	Precision	0.559	0.467	0.609	0.473
	Recall	0.518	0.468	0.615	0.472
	MR	0.088	0.085	0.078	0.092
	F1-Score	0.48	0.467	0.603	0.414
Caltech	Precision	0.439	0.472	0.58	0.403
	Recall	0.544	0.518	0.634	0.523
	MR	0.09	0.089	0.078	0.092
	F1-Score	0.449	0.49	0.595	0.419

[**Observation:** All the datasets have the maximum classification rate with ANN Classifier.]

TABLE 4.6

Classification Performances by Feature Extraction using Sauvola's Local Threshold Selection with Four Different Public Datasets using Four Different Classifiers

Datasets	Metrics	KNN	RIDOR	ANN	SVM
Wang	Precision	0.772	0.694	0.84	0.797
	Recall	0.765	0.695	0.84	0.798
	MR	0.054	0.067	0.042	0.05
	F1-Score	0.762	0.694	0.84	0.796
OT-Scene	Precision	0.61	0.619	0.761	0.632
	Recall	0.559	0.615	0.759	0.646
	MR	0.08	0.076	0.055	0.075
	F1-Score	0.57	0.615	0.759	0.633
Corel	Precision	0.712	0.641	0.841	0.757
	Recall	0.671	0.64	0.839	0.738
	MR	0.072	0.073	0.042	0.058
	F1-Score	0.649	0.64	0.837	0.732
Caltech	Precision	0.607	0.697	0.797	0.631
	Recall	0.647	0.697	0.809	0.692
	MR	0.079	0.059	0.04	0.073
	F1-Score	0.575	0.696	0.799	0.633

[**Observation:** All the datasets have highest classification rates with ANN Classifier.]

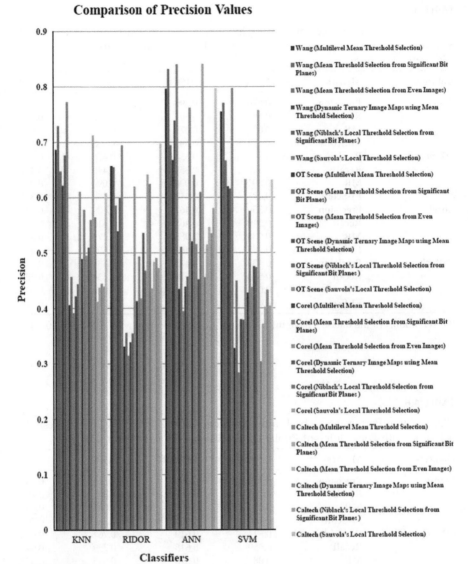

FIGURE 4.11
Comparison of Precision Values of Classification with Six Different Techniques of Feature Extraction for All Four Datasets using Four Different Classifiers.
[**Observation:** Classification by feature extraction with Sauvola's local threshold selection for binarization has the highest Precision rate among all the proposed techniques.]

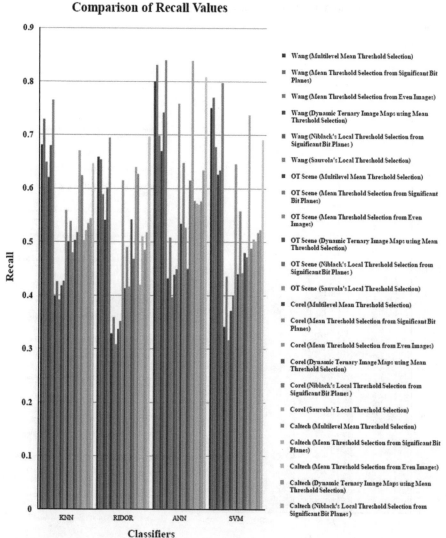

Comparison of Recall Values

Legend:

- Wang (Multilevel Mean Threshold Selection)
- Wang (Mean Threshold Selection from Significant Bit Planes)
- Wang (Mean Threshold Selection from Even Images)
- Wang (Dynamic Ternary Image Maps using Mean Threshold Selection)
- Wang (Niblack's Local Threshold Selection from Significant Bit Planes)
- Wang (Sauvola's Local Threshold Selection)
- OT Scene (Multilevel Mean Threshold Selection)
- OT Scene (Mean Threshold Selection from Significant Bit Planes)
- OT Scene (Mean Threshold Selection from Even Images)
- OT Scene (Dynamic Ternary Image Maps using Mean Threshold Selection)
- OT Scene (Niblack's Local Threshold Selection from Significant Bit Planes)
- OT Scene (Sauvola's Local Threshold Selection)
- Corel (Multilevel Mean Threshold Selection)
- Corel (Mean Threshold Selection from Significant Bit Planes)
- Corel (Mean Threshold Selection from Even Images)
- Corel (Dynamic Ternary Image Maps using Mean Threshold Selection)
- Corel (Niblack's Local Threshold Selection from Significant Bit Planes)
- Corel (Sauvola's Local Threshold Selection)
- Caltech (Multilevel Mean Threshold Selection)
- Caltech (Mean Threshold Selection from Significant Bit Planes)
- Caltech (Mean Threshold Selection from Even Images)
- Caltech (Dynamic Ternary Image Maps using Mean Threshold Selection)
- Caltech (Niblack's Local Threshold Selection from Significant Bit Planes)

FIGURE 4.12

Comparison of Recall Values of Classification with Six Different Techniques of Feature Extraction for All Four Datasets using Four Different Classifiers.

[**Observation:** Classification by feature extraction with Sauvola's local threshold selection for binarization has the highest Recall rate among all the proposed techniques.]

4.18 Statistical Significance

This chapter demonstrated a test to establish statistical significance of the enhanced classification results recorded with the proposed

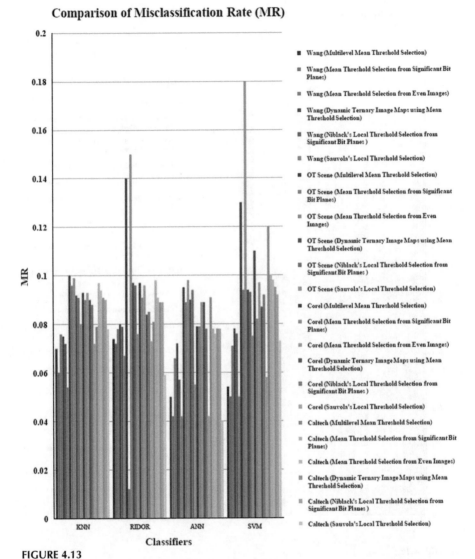

FIGURE 4.13
Comparison of MR with Six Different Techniques of Feature Extraction for All Four Datasets using Four Different Classifiers.
[**Observation:** Classification by feature extraction with Sauvola's local threshold selection for binarization has the lowest MR among all the proposed techniques.]

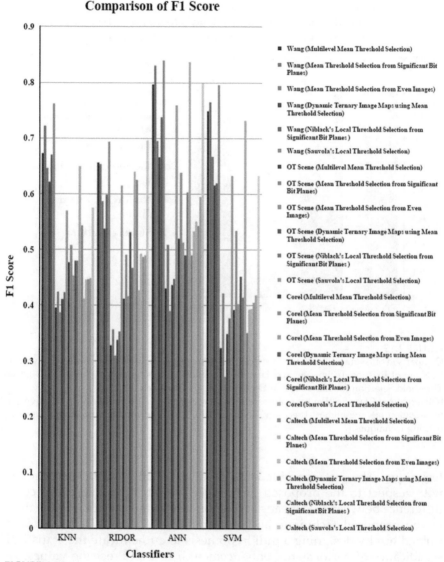

FIGURE 4.14
Comparison of F1-Score with Six Different Techniques of Feature Extraction for All Four Datasets using Four Different Classifiers.
[**Observation:** Classification by feature extraction with Sauvola's local threshold selection for binarization has the highest F1-Score among all the proposed techniques.]

Comparison fo Classification results by feature extraction using binarization with Sauvola's local threshold selection with respect to state-of-the art techniques of feature extraction

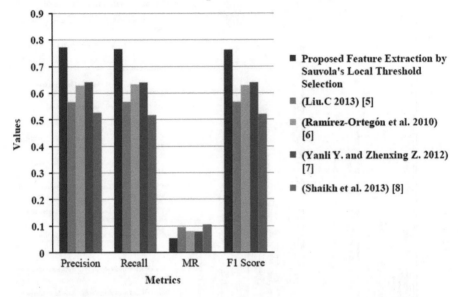

FIGURE 4.15
Comparison of Average Precision, Average Recall, Average Misclassification Rate and Average F1-Score of Existing Techniques wrt Feature Extraction by Binarization with Sauvola's Local Threshold Selection for Classification Done with KNN Classifier.
[**Observation:** Classification by feature extraction with Sauvola's local threshold selection for binarization has the highest Precision, Recall and F1-Score and the lowest MR compared to existing techniques.]

technique. The test computed the *p*-values of the existing techniques with respect to the proposed technique. The *t*-test revealed the actual difference between two means corresponding to the variation in the data of the proposed technique and the existing techniques. The test was carried out by designing a null hypothesis to authenticate the statistical significance of performance enhancement based on Precision values.

HYPOTHESIS 4.1: *There is no difference between the Precision values of classification result obtained by feature extraction using Sauvola's local threshold selection with respect to the existing techniques.*

The values of *mean, standard deviation (SD)* and *standard error of the mean (SEM)* of the Precision values of classification of different categories for feature extraction using BTC with color clumps is given in Table 4.7.

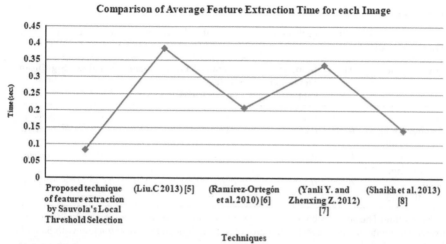

FIGURE 4.16

Comparison of Average Time Consumed for Feature Extraction.
[**Observation:** Feature extraction with Sauvola's local threshold selection for binarization has consumed minimum average time per image compared to the existing techniques.]

Table 4.8 gives the calculated values of t as t-calc and the p-values. The p-values generated after the comparison of Precision values for classification by feature extraction with the existing techniques with respect to the proposed technique have revealed a significant difference in Precision results for feature extraction with all the existing techniques. Therefore, the null hypothesis of equal Precision rates for the previously mentioned existing algorithm compared to the proposed algorithm is rejected for all the existing techniques. Further, the *mean, standard deviation* and *standard error of the mean (SEM)* for Precision values for classification by feature extraction with Sauvola's local threshold selection technique for feature extraction given in Table 4.7 is compared to that of the existing techniques in Table 4.8. It is observed that the mean Precision value for the novel technique has exceeded the mean Precision values of existing techniques. The value of *SD* and *SEM* for the novel

TABLE 4.7

Value of Mean Standard Deviation and Standard Error of the Mean for Precision Values of Classification by Feature Extraction using Sauvola's Local Threshold Selection for Binarization

Proposed Technique	Mean	Standard Deviation (SD)	Standard Error of the Mean (SEM)
Feature Extraction with Sauvola's Local Threshold Selection	0.783	0.117	0.039

TABLE 4.8

t-test for Evaluating Significance of Feature Extraction with Sauvola's Local Threshold Selection

Comparison	Mean	Standard Deviation (SD)	Standard Error of the Mean (SEM)	*t-calc*	*p*-Value	Significance
Liu, 2013	0.626	0.22	0.073	3.03	0.016	*Significant*
Ramírez-Ortegón et al., 2010	0.601	0.24	0.08	2.87	0.02	*Significant*
Yanli and Zhenxing, 2012	0.638	0.223	0.074	2.79	0.023	*Significant*
Shaikh et al., 2013	0.595	0.214	0.071	3.88	0.004	*Significant*

[**Observation:** The *p* values have indicated significant differences in Precision results for all the existing techniques with respect to the proposed technique of feature extraction with Sauvola's local threshold selection.]

technique is also less compared to the existing techniques. Therefore, it can be concluded that the introduced technique of feature extraction has greater consistency compared to the existing techniques. Hence, the proposed method has shown statistically significant contribution for enhanced classification compared to the existing techniques.

The *t-test* is performed with four different comparisons. Therefore, conduction of a post hoc analysis is necessary to accept the significance of the derived results using the Bonferroni correction. The value of α is considered to be *0.05* for the *t-test*. The expression to calculate the Bonferroni correction value follows:

$$Bonferroni\_\alpha = \frac{\alpha}{Number\_of\_tests}$$

Here, $\alpha = 0.05$

$$Number\_of\_tests = 4$$

Therefore,

$$Bonferroni\_\alpha = 0.013$$

Now, the *p-values* in Table 4.8 are compared to the newly calculated *Bonferroni_α* value of 0.013.

The comparison shows that the *p-value* calculated for evaluation of the proposed technique to the technique used by Shaikh [8] in Table 4.8 is smaller than *Bonferroni_α* value. Therefore, the two techniques have maintained statistical difference even after post-hoc analysis. However, the rest of the *p-values* are not smaller than *Bonferroni_α* value, although they have shown significant statistical difference during *t-test*.

Chapter Summary

The chapter carried out elaborate discussions on various feature extraction techniques based on binarization. It also displayed the usefulness of bit plane slicing for robust feature extraction. The propositions in the chapter are supported with MATLAB snippets for hands-on practice for the reader.

References

1. Kekre, H. B., Thepade, S., Das, R. K. and Ghosh, S., 2014. Multilevel block truncation coding with diverse color spaces for image classification. In International Conference on Advances in Technology and Engineering, ICATE, IEEE, pp. 1–7.
2. Kekre, H. B., Thepade, S., Das, R. K. and Ghosh, S., 2014. Performance boost of block truncation coding based image classification using bit plane slicing. *International Journal of Computer Applications*, 47(15): 36–41.
3. Thepade, S., Das, R. and Ghosh, S., 2013. Performance comparison of feature vector extraction techniques in rgb color space using block truncation coding for content based image classification with discrete classifiers. In India Conference, INDICON, Annual IEEE Digital Object Identifier: 10.1109/INDCON.2013.6726053 Publication, pp. 1–6.
4. Thepade, S., Das, R. and Ghosh, S., 2015. Content based image classification with thepade's static and dynamic ternary block truncation coding. *International Journal of Engineering Research* 4(1): 13–17.
5. Liu. C., 2013. A new finger vein feature extraction algorithm. In IEEE 6th International Congress on Image and Signal Processing, CISP, pp. 395–399.
6. Ramírez-Ortegón, M. A. and Rojas, R., 2010. Unsupervised evaluation methods based on local gray-intensity variances for binarization of historical documents. In Proceedings of the International Conference on Pattern Recognition, pp. 2029–2032.
7. Yanli, Y. and Zhenxing, Z., 2012. A novel local threshold binarization method for QR image. In IET International Conference on Automatic Control and Artificial Intelligence, ACAI, pp. 224–227.
8. Shaikh, S. H., Maiti, A. K. and Chaki, N., 2013. A new image binarization method using iterative partitioning. *Machine Vision and Applications*, 24(2): 337–350.

5

Content-Based Feature Extraction: Image Transforms

5.1 Prelude

Image transforms are essential to transfer the domain of an image for significant descriptor definition. Feature extraction from images in the frequency domain has often appeared to be beneficial compared to that of the spatial domain due to better visibility of patterns. Image transforms are considered as representations of images as a class of unitary matrices. A transform is useful in creating basis images for image representation by means of extrapolating the image to a set of basis functions. The process never alters the image content while shifting the image representation from spatial domain to frequency domain. Transforms convert spatial information to frequency domain information, where certain operations are easier to perform. For example, convolution operations can be reduced to matrix multiplication in the frequency domain. Two advantages are recognized for transform techniques. Firstly, the critical components of image patterns can be isolated by transformation, and those components can have direct access for analysis. Secondly, the transformation process has the ability to condense image data into a compact form, which has been helpful for radical reduction of feature vector size. This chapter implements these properties of image transforms to extract feature vectors for content-based image classification. Individual classification performances are measured with a full feature vector set generated by applying different transform techniques and with subsequent reduction of feature vector size toward the high frequency components at the upper portion of the image. Hence, the technique has identified the suitability of visual words method for generation of feature vectors. Visual words comprise small pieces of image data, which can carry significant information about the entire image. The energy compaction property of transform techniques have efficiently generated the features as visual words by concentrating important image information in a block of few transform components for each color component extracted from the image.

5.2 Generating Partial Energy Coefficient from Transformed Images

The significance of any image transform is to relocate the high-frequency components toward the upper end of the image and the low-frequency components toward the lower end of the image. The mentioned property is used as a tool to drastically reduce the size of feature vectors by exclusion of insignificant coefficients as shown in Fig. 5.1. A smaller dimension of feature vector has resulted in less time for comparison of feature vectors in the classification process.

Extraction of feature vectors is carried out in *14* steps from the transformed images. Initially, all coefficients of transformed images are considered as feature vectors that have included 100% of the transformed coefficients. Henceforth, gradual reduction in size of the feature vectors as *50%, 25%, 12.5%, 6.25%, 3.125%, 1.5625%, 0.7813%, 0.39%, 0.195%, 0.097%, 0.048%, 0.024%, 0.012%* and *0.06%* respectively of the complete transformed coefficients is carried out to locate the high-frequency component of the transformed image. For an image of dimension *256 × 256*, the process has reduced the feature size from *256 × 256* to *2 × 2* per color component.

The following steps are used for classification with feature extraction using partial energy coefficients:

- Red, Green and Blue color components are extracted from a given image.
- Image Transform is applied on each of the components to extract feature vectors.
- The extracted feature vectors from each of the components are stored as a complete set of feature vectors.
- Further, partial coefficients from the entire feature vector set are extracted to form the feature vector database.

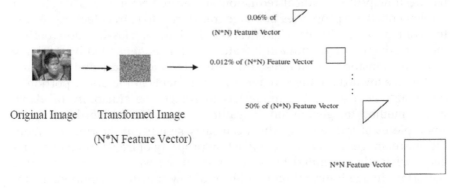

FIGURE 5.1
Steps for Extraction of Partial Energy Coefficient from Image.

TABLE 5.1

Dimension of Feature Vector for Different Percentages of Partial Energy Coefficients

Partial Coefficients (%)	Number of Elements in Feature Vector	Partial Coefficients (%)	Number of Elements in Feature Vector	Partial Coefficients (%)	Number of Elements in Feature Vector
100	65536	3.125	2048	0.097	64
50	32768	1.5625	1024	0.048	32
25	16384	0.7813	512	0.024	16
12.5	8192	0.39	256	0.012	8
6.25	4096	**0.195**	128	**0.006**	4

[**Observation:** Highest number of feature vectors elements has been noted with 100% transformed coefficients, and lowest number of feature vectors are observed with 0.006% of energy coefficients.]

- The feature vector database, with 100% transformed coefficients and partial coefficients ranging from 50% of the complete set of feature vectors till 0.06% of the complete set of feature vectors, is constructed as shown in Fig. 5.1.
- The feature vectors of the query image for the whole set of feature vectors and for partial coefficient of feature vectors are compared with the database images for classification results.
- Classifications are done with the entire training set for the query images and are compared for the highest classification result to find the best percentage of partial coefficient for feature extraction technique.
- Consider the best percentage thus inferred as the feature set extracted by applying image transform.

The size of feature vectors for each percentage of partial energy coefficients is given in Table 5.1.

The aforesaid steps are performed to extract features using partial energy coefficients by applying discrete cosine transform (DCT), Walsh-Hadamard transform, Kekre transform, discrete sine transform (DST) and Hartley transform on the database images.

5.3 Code Example (MATLAB®)

#Read an image

```
i=imread('path\imagefile');
i=imresize(i,[256 256]); #resizing the image
```

#Separate the color components
```
r=i(:,:,1); #red component
[m n p]=size(r); #dimension of red component
g=i(:,:,2); #green component
[x y z]=size(g); #dimension of green component
b=i(:,:,3); #blue component
[j k l]=size(b); #dimension of blue component
```

#Apply Transform
```
r1=dst(r); #applying discrete sine transform (dst) on red component
g1=dst(g); #applying discrete sine transform (dst) on green
component
b1=dst(b); #applying discrete sine transform (dst) on blue
component
```

#Extract Partial Transform Coefficients
```
count=0;
for a=1:m/64
for b=1:n/64
if (a+b)<=5
count=count+1;
r2(count)=r1(a,b);
g2(count)=g1(a,b);
b2(count)=b1(a,b);
end
end
```

5.4 Coding Exercise

The previous code example is for generating partial transform coefficients. Refer to the example to complete the following assignments:

- *What percentage of fractional coefficients are extracted in the previous example?*
- *Apply DCT, Walsh and Kekre transforms to extract all proportions of partial coefficients.*
- *Test classification results for each variety.*

5.5 Computational Complexity for the Image Transforms

This chapter has discussed five different image transforms applied in the research work for feature vector extraction. The comparative computational complexity of the transforms is displayed in Table 5.2 for an image of dimension $N \times N$.

TABLE 5.2

Comparison of Computational Complexity of Applied Transforms

Transforms	Number of Additions For $N \times N$ image	Number of Multiplications	Total Additions for Transform of 256×256 Image Dimension	Total Computations Comparison (For 256×256 Image Dimension)
DCT	$2N^2(N-1)$	$N^2(2N)$	301858816	100
Walsh	$2N^2(N-1)$	0	33423360	11.07
Kekre	$N[N(N+1) - 2]$	$2N(N-2)$	17882624	5.92
DST	$2N^2(N-1)$	$N^2(2N)$	301858816	100
Hartley	$2N^2(N-1)$	$N^2(2N)$	301858816	100

[**Observation:** Here one multiplication has been considered as eight additions for the second-last row computations, and DCT computations have been considered to be 100% for comparison in the last row.]

5.6 Feature Extraction with Partial Energy Coefficient

Five different image transforms—namely, DCT, Walsh transform, Kekre transform, DST and Hartley transform—have been applied on the test images to extract features with partial energy coefficient for content-based image classification. The classification results are evaluated by a 10-fold cross-validation scheme. Each of the techniques has been described in the following subsections and the classification results with various percentages of partial energy have been compared with four different metrics—namely, Precision, Recall, Misclassification Rate (MR) and F1-Score. Four different classifiers—namely, K Nearest Neighbor (KNN), Ripple-Down Rule (RIDOR), Artificial Neural Network (ANN) and Support Vector Machine (SVM)—are used for appraising the classification of four widely used public datasets, namely, Wang dataset (1000 images and 10 categories), Oliva Torralba (OT-Scene) dataset (2688 images and 8 categories), Corel dataset (10,800 images and 80 categories) and Caltech dataset (8127 images and 100 categories). defined The classifiers and the datasets have been described in Chapter 1. Initially, feature extraction with all the percentages of partial coefficients of all the considered transforms are assessed for MR and F1-Score values of classification. Test images from the Wang dataset are utilized for this purpose and the similarity measure of images is carried out with Mean Squared Error (MSE) distance measure.

5.6.1 Discrete Cosine Transform

Discrete cosine transform (DCT) has a close link to the discrete Fourier transform (DFT) and has been defined as a separable linear transformation process [1]. The working formula for two-dimensional DCT for application on an input image P to produce an output image Q can be demonstrated as in equation 5.1.

$$Q_{(i,j)} = a_i a_j \sum\sum A_{(u,v)} \cos \frac{\pi(2u+1)i}{2A} \cos \frac{\pi(2v+1)j}{2B} \tag{5.1}$$

$$a_1 = \begin{cases} 1/\sqrt{A}, & i = 0 \\ \sqrt{2/A}, & 1 \le i \le A - 1 \end{cases}$$

$$a_i = \begin{cases} 1/\sqrt{B}, & j = 0 \\ \sqrt{2/B}, & 1 \le j \le B - 1 \end{cases}$$

A and B are the size of rows and columns respectively for the considered image P.

The number of multiplications and additions required are $N^2(2N)$ and $N^2(2N - 2)$ respectively for a full 2-dimensional DCT for an $N \times N$ image.

The different percentages of fractional coefficients are evaluated for classification results by comparing the MR and F1-Score with the Wang dataset initially as shown in Fig. 5.2.

Fig. 5.2 has revealed maximum classification results with 0.012% of full feature size of transformed DCT coefficients and minimum with 100% of the coefficients. A gradual increase in F1-Score and decrease in MR is observed up to 0.012% of feature size. The MR increased and F1-Score decreased with a further percentage of 0.006% of partial coefficient. Further, the Precision, Recall, MR and F1-Score values for four different public datasets using four different classifiers—namely, KNN, RIDOR, ANN and SVM are given in Table 5.3.

5.6.2 Walsh Transform

A representation of Walsh transform matrix is done with a set of N rows denoted by W_j as shown in Algorithm 5.1. The matrix takes on the values of +1 and −1. The value of $W_j[0]$ has been considered 1 for all j. The definition of Walsh transform matrix is represented by a Hadamard matrix of order N. The Walsh code index has specified the Walsh transform matrix row as the row of the Hadamard matrix, which should be an integer in the range $[0, ..., N - 1]$. The corresponding Hadamard output code for the Walsh code index equal to an integer j, possessed specifically j zero crossings, for $j = 0, 1, ..., N - 1$. The partial coefficients of feature vectors are generated from the transformed coefficients of Walsh transform [2].

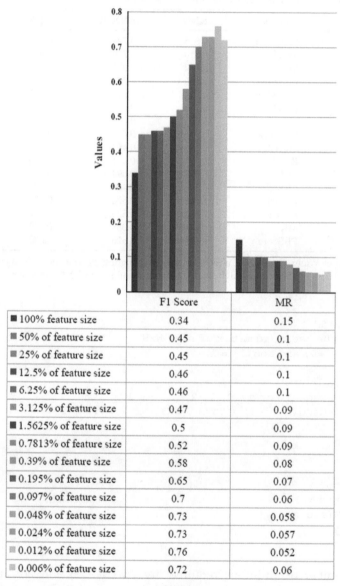

Comparison of Average F1 Score and MR for feature extraction with different partial coefficients of DCT

	F1 Score	MR
■ 100% feature size	0.34	0.15
■ 50% of feature size	0.45	0.1
■ 25% of feature size	0.45	0.1
■ 12.5% of feature size	0.46	0.1
■ 6.25% of feature size	0.46	0.1
■ 3.125% of feature size	0.47	0.09
■ 1.5625% of feature size	0.5	0.09
■ 0.7813% of feature size	0.52	0.09
■ 0.39% of feature size	0.58	0.08
■ 0.195% of feature size	0.65	0.07
■ 0.097% of feature size	0.7	0.06
■ 0.048% of feature size	0.73	0.058
■ 0.024% of feature size	0.73	0.057
■ 0.012% of feature size	0.76	0.052
■ 0.006% of feature size	0.72	0.06

Metrics

FIGURE 5.2

F1-Score and MR for Classification by Feature Extraction with Partial DCT Coefficients. [**Observation:** Maximum F1-Score and minimum MR with 0.012% of full feature size of DCT coefficients.]

TABLE 5.3

Classification Performance with Partial DCT Coefficients with Four Different Public Datasets using Four Different Classifiers

Datasets	Metrics	KNN	RIDOR	ANN	SVM
Wang	Precision	0.825	0.794	0.941	0.858
	Recall	0.818	0.790	0.941	0.851
	MR	0.052	0.053	0.012	0.020
	F1-Score	0.760	0.750	0.941	0.851
OT-Scene	Precision	0.924	0.612	0.743	0.619
	Recall	0.923	0.610	0.739	0.622
	MR	0.014	0.073	0.055	0.070
	F1-Score	0.923	0.611	0.739	0.619
Corel	Precision	0.623	0.462	0.671	0.651
	Recall	0.598	0.461	0.680	0.634
	MR	0.077	0.094	0.069	0.071
	F1-Score	0.571	0.460	0.672	0.613
Caltech	Precision	0.615	0.526	0.677	0.559
	Recall	0.631	0.526	0.687	0.621
	MR	0.070	0.083	0.063	0.077
	F1-Score	0.619	0.526	0.678	0.570

[**Observations:** Wang, Corel and Caltech datasets have highest classification results with ANN Classifier; OT-Scene dataset has highest classification results with KNN classifier.]

ALGORITHM 5.1:
```
Let H be the hadamard matix of size NxN
Let seq(0)=0; and seq(1) = 1
Let N1=2
For k=1 to log2(N) -1
Cseq(n) = seq(n) for n = 0, 1, 2, ...., N1-1.
Cseq(n) = Cseq(n-2i+1) + 1 for n =N1, N1+1,....., 2N1-1
and i=0,1,2,......, N1-1.
seq=Cseq
N1=2N1.
End for
For n=0 to N-1
For j = 0 to N-1
W(n,j)=H(seq(n),j)
End for j
End for n
```

The number of additions required are $2N^2(N-1)$ and absolutely no multiplications are needed for the full 2-dimensional Walsh transform applied to an image of size $N \times N$. MR and F1-Score of different percentages of partial energy coefficients of Walsh transform are compared with the Wang dataset as shown in Fig. 5.3.

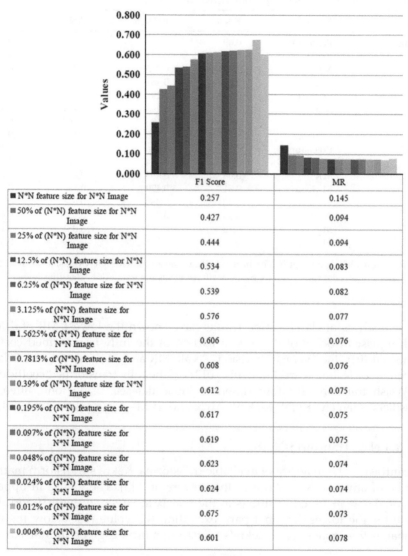

Comparison of Average F1 Score and MR for feature extraction with different partial coefficients of Walsh Transform

	F1 Score	MR
■ N*N feature size for N*N Image	0.257	0.145
■ 50% of (N*N) feature size for N*N Image	0.427	0.094
■ 25% of (N*N) feature size for N*N Image	0.444	0.094
■ 12.5% of (N*N) feature size for N*N Image	0.534	0.083
■ 6.25% of (N*N) feature size for N*N Image	0.539	0.082
■ 3.125% of (N*N) feature size for N*N Image	0.576	0.077
■ 1.5625% of (N*N) feature size for N*N Image	0.606	0.076
■ 0.7813% of (N*N) feature size for N*N Image	0.608	0.076
■ 0.39% of (N*N) feature size for N*N Image	0.612	0.075
■ 0.195% of (N*N) feature size for N*N Image	0.617	0.075
■ 0.097% of (N*N) feature size for N*N Image	0.619	0.075
■ 0.048% of (N*N) feature size for N*N Image	0.623	0.074
■ 0.024% of (N*N) feature size for N*N Image	0.624	0.074
■ 0.012% of (N*N) feature size for N*N Image	0.675	0.073
■ 0.006% of (N*N) feature size for N*N Image	0.601	0.078

Metrics

FIGURE 5.3

F1-Score and MR for Classification by Feature Extraction with Partial Walsh Transform Coefficients.

[**Observation:** Maximum F1-Score and minimum MR with 0.012% of full feature size of Walsh transform coefficients.]

TABLE 5.4

Classification Performance with Partial Walsh Transform Coefficients with Four Different Public Datasets using Four Different Classifiers

Datasets	Metrics	KNN	RIDOR	ANN	SVM
Wang	Precision	0.69	0.433	0.544	0.517
	Recall	0.68	0.431	0.547	0.532
	MR	0.064	0.095	0.079	0.084
	F1-Score	0.675	0.431	0.544	0.515
OT-Scene	Precision	0.388	0.298	0.451	0.398
	Recall	0.388	0.298	0.451	0.408
	MR	0.12	0.13	0.095	0.12
	F1-Score	0.381	0.298	0.449	0.381
Corel	Precision	0.53	0.476	0.554	0.482
	Recall	0.54	0.484	0.572	0.498
	MR	0.08	0.093	0.078	0.087
	F1-Score	0.517	0.479	0.557	0.457
Caltech	Precision	0.401	0.369	0.518	0.309
	Recall	0.462	0.36	0.559	0.441
	MR	0.12	0.14	0.081	0.15
	F1-Score	0.381	0.364	0.534	0.322

[Observations: Wang dataset has highest classification results with KNN classifier; OT-Scene, Corel and Caltech datasets have highest classification results with ANN classifier.]

The analysis of classification results have shown the lowest MR of 0.064 and highest F1-Score of 0.675 with 0.012% of the fully transformed coefficients. Further, the average Precision, Recall, MR and F1-Score values of classification for feature extraction with 0.012% of the fully transformed coefficient of Walsh transform for four different public datasets using four different classifiers—namely, KNN, RIDOR, ANN and SVM—are given in Table 5.4.

5.6.3 Kekre Transform

In contrast to the majority of available transforms, Kekre's transform matrix can be of any size $N \times N$ not in the powers of 2 [3]. The values in upper-diagonal and diagonal positions in the matrix are assigned with 1 and the value for the lower-diagonal part, excluding the value just below the diagonal, is 0. A generalized Kekre's transform matrix is given in Matrix. 5.1.

$$K_{NXN} = \begin{bmatrix} 1 & 1 & 1 & .. & 1 & 1 \\ -N+1 & 1 & 1 & .. & 1 & 1 \\ 0 & -N+2 & 1 & .. & 1 & 1 \\ \vdots & \vdots & \vdots & : & \vdots & \vdots \\ 0 & 0 & 0 & .. & 1 & 1 \\ 0 & 0 & 0 & .. & -N+(N-1) & 1 \end{bmatrix}$$

TABLE 5.5

Classification Performance with Partial Kekre Transform Coefficients with Four Different Public Datasets using Four Different Classifiers

Datasets	Metrics	KNN	RIDOR	ANN	SVM
Wang	Precision	0.54	0.413	0.53	0.448
	Recall	0.541	0.415	0.536	0.5
	MR	0.086	0.097	0.087	0.092
	F1-Score	0.541	0.413	0.531	0.468
OT-Scene	Precision	0.377	0.289	0.434	0.381
	Recall	0.381	0.289	0.433	0.38
	MR	0.14	0.16	0.097	0.13
	F1-Score	0.377	0.289	0.434	0.381
Corel	Precision	0.516	0.449	0.54	0.464
	Recall	0.526	0.455	0.56	0.488
	MR	0.1	0.099	0.081	0.099
	F1-Score	0.411	0.45	0.544	0.437
Caltech	Precision	0.455	0.419	0.523	0.391
	Recall	0.514	0.459	0.552	0.528
	MR	0.094	0.095	0.08	0.1
	F1-Score	0.443	0.437	0.554	0.43

[Observations: The Wang dataset has highest classification results with KNN classifier; OT-Scene, Corel and Caltech datasets have highest classification results with ANN classifier.]

The required number of multiplications are *2N(N-2)* and the required number of additions are $N(N^2+N - 2)$ for taking Kekre's transform of an $N \times N$ image.

The term $K(x, y)$ of Kekre transform matrix can be generated by the expression given in equation 5.2.

$$K(x, y) = \begin{cases} 1, & x \leq y \\ -N + (x - 1), & x = y + 1 \\ 0, & x > y + 1 \end{cases} \qquad (5.2)$$

Partial coefficients are extracted as feature vectors from the transformed elements of the images. A comparison of F1-Score and MR for feature extraction with different percentages of partial coefficients is given in Fig. 5.4.

The comparison in Fig. 5.4 has established the highest F1-Score and lowest MR for classification with 12.5% of the complete transformed coefficients obtained by using Kekre transform. Henceforth, the average Precision, Recall, MR and F1-Score values of classification for feature extraction with 12.5% of the fully transformed coefficient of Kekre transform for four different public datasets using four different classifiers—namely, KNN, RIDOR, ANN and SVM—have been given in Table 5.5.

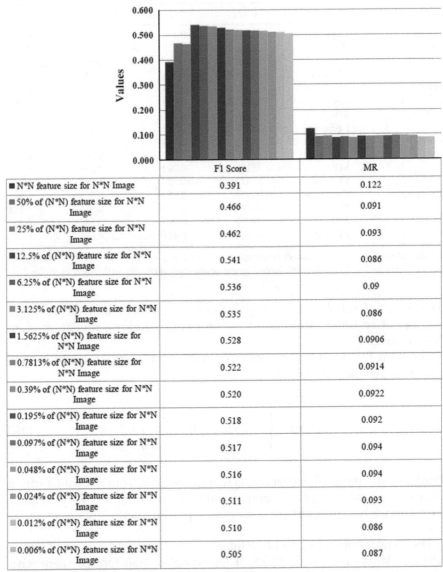

Comparison of Average F1 Score and MR for feature extraction with different partial coefficients of Kekre Transform

	F1 Score	MR
∎ N*N feature size for N*N Image	0.391	0.122
∎ 50% of (N*N) feature size for N*N Image	0.466	0.091
∎ 25% of (N*N) feature size for N*N Image	0.462	0.093
∎ 12.5% of (N*N) feature size for N*N Image	0.541	0.086
∎ 6.25% of (N*N) feature size for N*N Image	0.536	0.09
∎ 3.125% of (N*N) feature size for N*N Image	0.535	0.086
∎ 1.5625% of (N*N) feature size for N*N Image	0.528	0.0906
∎ 0.7813% of (N*N) feature size for N*N Image	0.522	0.0914
∎ 0.39% of (N*N) feature size for N*N Image	0.520	0.0922
∎ 0.195% of (N*N) feature size for N*N Image	0.518	0.092
∎ 0.097% of (N*N) feature size for N*N Image	0.517	0.094
∎ 0.048% of (N*N) feature size for N*N Image	0.516	0.094
∎ 0.024% of (N*N) feature size for N*N Image	0.511	0.093
∎ 0.012% of (N*N) feature size for N*N Image	0.510	0.086
∎ 0.006% of (N*N) feature size for N*N Image	0.505	0.087

Metrics

FIGURE 5.4
F1-Score and MR for Classification by Feature Extraction with Partial Kekre Transform Coefficients.
[**Observation:** Maximum F1-Score and minimum MR are reached with 12.5% of full feature size of Kekre transform coefficients.]

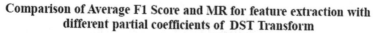

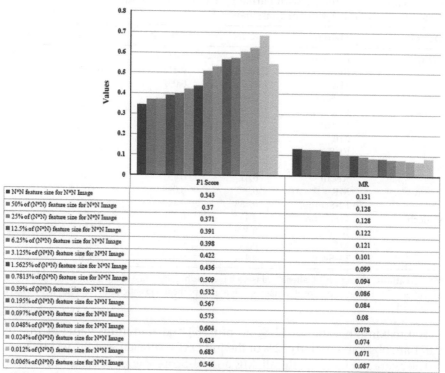

	F1 Score	MR
■ N*N feature size for N*N Image	0.343	0.131
■ 50% of (N*N) feature size for N*N Image	0.37	0.128
■ 25% of (N*N) feature size for N*N Image	0.371	0.128
■ 12.5% of (N*N) feature size for N*N Image	0.391	0.122
■ 6.25% of (N*N) feature size for N*N Image	0.398	0.121
■ 3.125% of (N*N) feature size for N*N Image	0.422	0.101
■ 1.5625% of (N*N) feature size for N*N Image	0.436	0.099
■ 0.7813% of (N*N) feature size for N*N Image	0.509	0.094
■ 0.39% of (N*N) feature size for N*N Image	0.532	0.086
■ 0.195% of (N*N) feature size for N*N Image	0.567	0.084
■ 0.097% of (N*N) feature size for N*N Image	0.573	0.08
■ 0.048% of (N*N) feature size for N*N Image	0.604	0.078
■ 0.024% of (N*N) feature size for N*N Image	0.624	0.074
■ 0.012% of (N*N) feature size for N*N Image	0.683	0.071
■ 0.006% of (N*N) feature size for N*N Image	0.546	0.087

Metrics

FIGURE 5.5
F1-Score and MR for Classification by Feature Extraction with Partial DST Coefficients.
[**Observation:** Maximum F1-Score and minimum MR is reached with 0.012% of full feature size of DST coefficients.]

5.6.4 Discrete Sine Transform

Discrete sine transform (DST) is defined as a Fourier-related transform that utilizes a purely real matrix [3]. It has been described by a $N \times N$ sine transform matrix and is considered as a linear and invertible function. Generation of the DST matrix can be carried out by row in the arrangement of the sequences given in equation 5.3.

$$y(i, j) = \sqrt{2/(N+1)} \sin [\Pi (i+1)(j+1)/(N+1)] \tag{5.3}$$

for $0 \leq i, j <= N - 1$

Partial coefficients of DST are extracted as feature vectors from images in the Wang dataset. A comparison of F1-Score and MR for feature extraction with different percentages of partial coefficients of DST is given in Fig. 5.5.

TABLE 5.6

Classification Performance with Partial DST Coefficients with Four Different Public Datasets using Four Different Classifiers

Datasets	Metrics	KNN	RIDOR	ANN	SVM
Wang	Precision	0.67	0.41	0.557	0.49
	Recall	0.68	0.414	0.562	0.497
	MR	0.071	0.097	0.08	0.092
	F1-Score	0.683	0.412	0.558	0.466
OT-Scene	Precision	0.376	0.31	0.437	0.354
	Recall	0.379	0.313	0.438	0.36
	MR	0.14	0.16	0.097	0.15
	F1-Score	0.372	0.31	0.436	0.329
Corel	Precision	0.501	0.446	0.544	0.449
	Recall	0.51	0.45	0.556	0.471
	MR	0.097	0.098	0.09	0.1
	F1-Score	0.488	0.447	0.542	0.43
Caltech	Precision	0.45	0.406	0.424	0.327
	Recall	0.499	0.431	0.512	0.492
	MR	0.1	0.12	0.097	0.14
	F1-Score	0.424	0.417	0.456	0.372

[Observations: Wang and Caltech datasets have the highest classification results with KNN classifier; OT-Scene and Corel datasets have the highest classification results with ANN classifier.]

The maximum value of F1-Score of 0.683 and minimum value of MR of 0.071 are observed with 0.012% of the full DST coefficients of the test images.

The average Precision, Recall, MR and F1-Score values of classification for feature extraction with 0.012% of the fully transformed coefficient of DST for four different public datasets using four different classifiers—namely, KNN, RIDOR, ANN and SVM are given in Table 5.6.

5.6.5 Discrete Hartley Transform

Discrete Hartley transform (DHT) is termed as a Fourier-related transform of discrete periodic data, which transforms the real inputs to real outputs while avoiding the elementary association of complex numbers [4]. It has acted as a linear operator and has been the discrete analogue of the continuous Hartley Transform. DHT has been a linear, invertible function $H: R^n \rightarrow R^n$ (where R denotes the set of real numbers). Transformation of n real numbers starting from $x_0,..., x_{n-1}$ into n real numbers $h_0, ..., h_{n-1}$ can be performed by equation 5.4

$$H_k = \sum_{n=0}^{N-1} x_n \left[\cos\left(\frac{2\Pi}{N}nk\right) + \sin\left(\frac{2\Pi}{N}nk\right) \right]_{K=0,... ..., N-1} \qquad (5.4)$$

Interpretation of the transform has been done as the multiplication of the vector $(x_0, ..., x_{N-1})$ by an $N \times N$ matrix; therefore, the discrete Hartley

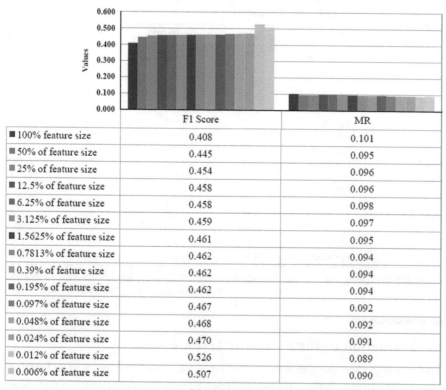

Comparison of Average F1 Score and MR for feature extraction with different partial coefficients of DHT

	F1 Score	MR
■ 100% feature size	0.408	0.101
■ 50% of feature size	0.445	0.095
■ 25% of feature size	0.454	0.096
■ 12.5% of feature size	0.458	0.096
■ 6.25% of feature size	0.458	0.098
■ 3.125% of feature size	0.459	0.097
■ 1.5625% of feature size	0.461	0.095
■ 0.7813% of feature size	0.462	0.094
■ 0.39% of feature size	0.462	0.094
■ 0.195% of feature size	0.462	0.094
■ 0.097% of feature size	0.467	0.092
■ 0.048% of feature size	0.468	0.092
■ 0.024% of feature size	0.470	0.091
■ 0.012% of feature size	0.526	0.089
■ 0.006% of feature size	0.507	0.090

Metrics

FIGURE 5.6
F1-Score and MR for Classification by Feature Extraction with Partial DHT Coefficients. [**Observation:** Maximum F1-Score and minimum MR are reached with 0.012% of full feature size of DHT coefficients.]

transform is designated as a linear operator. DHT of H_k multiplied by $1/N$ has facilitated the recovery of x_n from H_k. Hence, the matrix has been termed as invertible. Therefore, up to an overall scale factor, DHT is its own inverse. The required number of multiplications are $2N(N - 2)$ and additions are N $(N^2 + N - 2)$ for taking Hartley transform of an $N \times N$ image.

Feature vectors from test images of the Wang dataset are extracted as partial coefficients of DHT. A comparison of F1-Score and MR for feature extraction with different percentages of partial coefficients of DHT is given in Fig. 5.6.

The average Precision, Recall, MR and F1-Score values of classification for feature extraction with 0.012% of the fully transformed coefficient of DHT for four different public datasets using four different classifiers—namely, KNN, RIDOR, ANN and SVM are given in Table 5.7.

TABLE 5.7

Classification Performance with Partial DST Coefficients with Four Different Public Datasets using Four Different Classifiers

Datasets	Metrics	KNN	RIDOR	ANN	SVM
Wang	Precision	0.554	0.412	0.551	0.476
	Recall	0.553	0.416	0.55	0.501
	MR	0.07	0.09	0.086	0.09
	F1-Score	0.553	0.413	0.55	0.479
OT-Scene	Precision	0.371	0.298	0.362	0.357
	Recall	0.376	0.299	0.361	0.372
	MR	0.14	0.16	0.15	0.15
	F1-Score	0.373	0.298	0.361	0.347
Corel	Precision	0.171	0.121	0.221	0.152
	Recall	0.201	0.176	0.22	0.151
	MR	0.24	0.25	0.22	0.2
	F1-Score	0.18	0.121	0.221	0.15
Caltech	Precision	0.456	0.423	0.443	0.385
	Recall	0.522	0.456	0.532	0.52
	MR	0.098	0.095	0.09	0.097
	F1-Score	0.441	0.438	0.475	0.413

[Observations: Wang and Caltech datasets have highest classification results with KNN classifier; OT-Scene and Corel datasets have highest classification results with ANN classifier.]

5.7 Evaluation of the Proposed Techniques

The classification performances shown by each of the feature extraction techniques using partial transform coefficients are compared across four different classifiers and four public datasets as shown in Figs. 5.7–5.10.

The illustrations in Figs. 5.7–5.10 clearly reveal that classification results by feature extraction with partial DCT coefficients has outclassed all of the other four transforms—namely, Walsh transform, Kekre transform, DST and DHT.

The reason for the superior performance of DCT is due to the fact that the mean of the gray values being transformed is the first coefficient of DCT. Thus, the first coefficient of each block is the average tone of the participating gray values, which allows the subsequent coefficients to add an increasing number of details, facilitating better feature extraction. On the other hand, the $sin\theta$-based transforms get initiated with offset of 0.5 or 1, which results in a gentle mound in the first coefficient. This is not suitable for robust feature extraction from the given test images.

The number of iterations carried out by the transform techniques for feature vector calculation is observed to consume linear time. Each transform takes L number of iteration for calculating the coefficients. Thus for three color components (Red, Green and Blue), a total of $3L$ iterations are required. However, the selected partial coefficient is of $n\%$ (where $n = 50\%$ to 0.006%) of the whole transform coefficient generated. Consequently, the number of iterations required

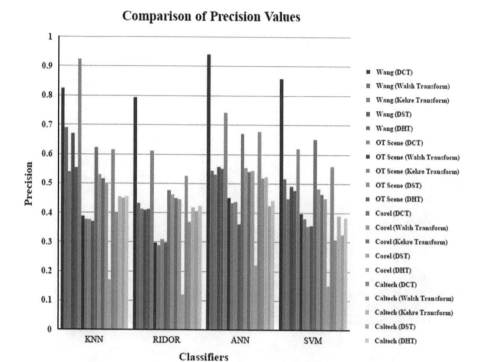

FIGURE 5.7
Comparison of Precision Values of Classification with Partial Coefficients of DCT, Walsh transform, Kekre transform, DST and DHT Method of Feature Extraction for All Four Datasets using Four Different Classifiers.
[**Observation:** Highest Precision value has been shown by feature extraction with partial DCT coefficients.]

for computing feature vectors with $n\%$ of transform coefficients is much less than $3L$. Thus, the computation for feature extraction is reduced to $n\%$ of $3L$. Thus, the total number of iterations can be given by linear time $O\ (3*n*N)$.

5.8 Comparison with Existing Techniques

It is inferred from the results in Section 5.4, that classification results for feature extraction with partial DCT coefficients have outshined the other proposed techniques.

Therefore, the classification results obtained by feature extraction with partial DCT coefficients (observed to be the best technique in Section 5.4) are compared with respect to state-of-the-art techniques [5–8] in Fig. 5.11. The existing techniques are tested with the Wang dataset and KNN classifier. Hence, the comparison with the proposed technique is carried out under the same environment.

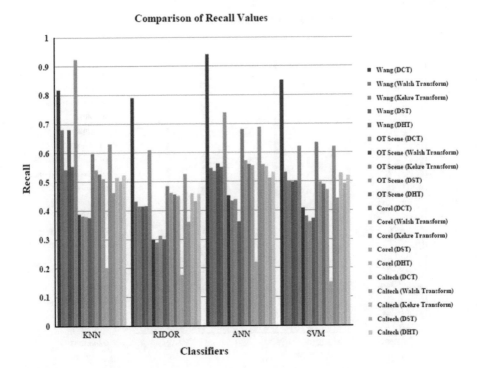

Comparison of Recall Values

Legend:
- Wang (DCT)
- Wang (Walsh Transform)
- Wang (Kekre Transform)
- Wang (DST)
- Wang (DHT)
- OT Scene (DCT)
- OT Scene (Walsh Transform)
- OT Scene (Kekre Transform)
- OT Scene (DST)
- OT Scene (DHT)
- Corel (DCT)
- Corel (Walsh Transform)
- Corel (Kekre Transform)
- Corel (DST)
- Corel (DHT)
- Caltech (DCT)
- Caltech (Walsh Transform)
- Caltech (Kekre Transform)
- Caltech (DST)
- Caltech (DHT)

Classifiers: KNN, RIDOR, ANN, SVM

FIGURE 5.8

Comparison of Recall Values of Classification with Partial Coefficients of DCT, Walsh Transform, Kekre Transform, DST and DHT Method of Feature Extraction for All the Four Datasets using Four Different Classifiers.

[**Observation:** Highest Recall value has been shown by feature extraction with partial DCT coefficients.]

The results in Fig. 5.11 have clearly established the supremacy of the proposed technique over the existing techniques.

Finally, the average feature extraction time from each image with the proposed feature extraction technique with partial DCT coefficients has been compared to the existing technique, as shown in Fig. 5.12.

The illustration shown in Fig. 5.12 has clearly established that the proposed technique of feature extraction with partial DCT coefficients has consumed much less time compared to the benchmarked technique [5–8].

5.9 Statistical Significance

A null hypothesis has been designed in *Hypothesis 5.1* to validate the statistical significance of performance improvement based on Precision values.

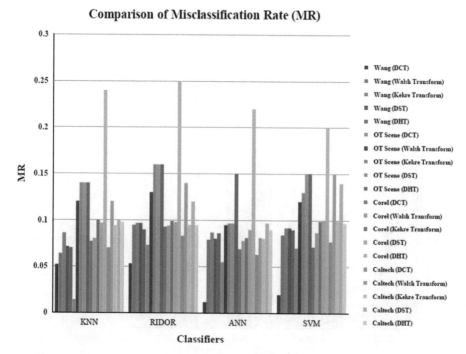

FIGURE 5.9
Comparison of MR of Classification with Partial Coefficients of DCT, Walsh Transform, Kekre Transform, DST and DHT Method of Feature Extraction for All Four Datasets using Four Different Classifiers.

[**Observation:** Lowest MR has been shown by feature extraction with partial DCT coefficients.]

HYPOTHESIS 1: *There is no significant difference between the Precision values of classification result obtained by feature extraction partial DCT coefficients with respect to the existing techniques.*

A paired *t*-test is conducted in Table 5.9 to determine if the difference in Precision values for classification is generated from a population with zero mean.

Initially, SD and SEM of the Precision values of classification of different categories for feature extraction with partial DCT coefficients has been given in Table 5.8.

The value of *t*-calc is computed by taking the difference between two sample means considered for each comparison in Table 5.9. The strength of evidence against the null hypothesis is measured by the *p*-value.

The *p*-values calculated have indicated significant difference in precision results for classification with the proposed technique of feature extraction, and the null hypothesis is rejected. The values of *mean, SD* and *SEM* of the existing techniques given in Table 5.8 are further compared to that of the proposed method in Table 5.9. It is observed that feature extraction with

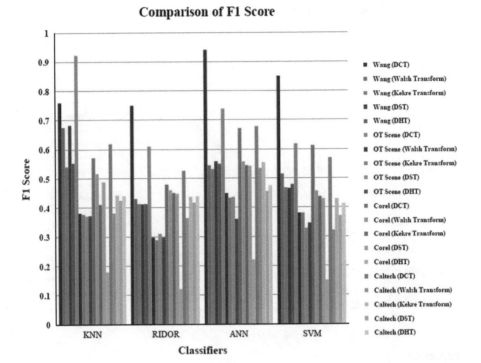

FIGURE 5.10
Comparison of F1-Score of Classification with Partial Coefficients of DCT, Walsh Transform, Kekre Transform, DST and DHT Method of Feature Extraction for All Four Datasets using Four Different Classifiers.
[**Observation:** Highest classification performance has been shown by feature extraction with partial DCT coefficients.]

the partial DCT coefficients method has a higher *mean* of Precision values for classification with less *SD* and *SEM* compared to all the existing techniques. Hence, it can be concluded that the proposed method has not only shown higher Precision values, but also has consistent performance because it has less *SD* and *SEM* compared to the existing techniques.

Therefore, the feature extraction technique with 0.012% of DCT coefficients has contributed significantly in boosting up the classification performance.

The *t-test* is performed with four different comparisons; a *post-hoc* analysis is necessary to accept the significance of the derived results using the *Bonferroni Correction*. The value of α is considered to be *0.05* for the *t-test*. The following expression calculates the *Bonferroni correction* value:

$$Bonferroni\_\alpha = \frac{\alpha}{Number\_of\_tests}$$

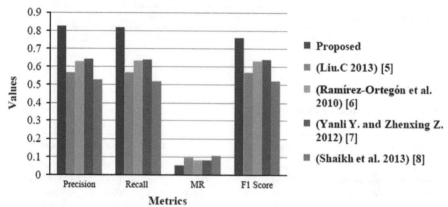

FIGURE 5.11

Comparison of Average Precision, Average Recall, Average MR and average F1-Score of Existing Techniques wrt Proposed Technique of Feature Extraction for Classification Done with KNN Classifier.

[**Observation:** Proposed technique has the highest Precision, Recall and F1-Score values as well as the lowest MR value compared to the existing techniques.]

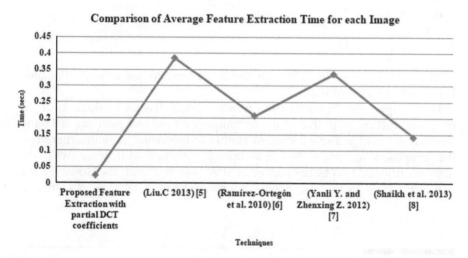

FIGURE 5.12

Comparison of Average Time Consumed for Feature Extraction.

[**Observation:** Feature extraction with partial DCT coefficients has consumed the minimum average time per image among all the proposed techniques.]

TABLE 5.8

Value of Mean, Standard Deviation and Standard Error of the Mean for Precision Values of Classification by Feature Extraction with Partial DCT Coefficients

Proposed Technique	Mean	SD	SEM
Feature extraction with partial DCT coefficients	0.801	0.145	0.071

TABLE 5.9

t test for Comparison of Precision Results of Feature Extraction with Partial DCT Coefficients

Comparison	Mean	SD	SEM	*t-calc*	*p*-Value	Significance
Liu, 2013	0.626	0.22	0.073	3.78	0.005	Significant
Ramírez-Ortegón et al., 2010	0.601	0.24	0.08	3.88	0.005	Significant
Yanli. and Zhenxing, 2012	0.638	0.223	0.074	3.23	0.012	Significant
Shaikh et al., 2013	0.595	0.214	0.071	4.26	0.003	Significant

[**Observation:** The *p* values have indicated significant differences in Precision results for classification for feature extraction using existing techniques compared to feature extraction using partial DCT coefficients.]

where $\alpha = 0.05$
Number_of_tests = 4
Therefore,
*Bonferroni_*$\alpha = 0.013$

Now, the *p-values* in Table 5.9 are compared to the newly calculated *Bonferroni_*α value of 0.013.

The comparison shows that the *p-values* calculated for evaluation of the proposed technique to that of the technique of *Yanli and Zhenxing 2012* and *Shaikh et al., 2013* in Table 5.9 are smaller than *Bonferroni_*α value. Therefore, the proposed technique has maintained statistical difference with the abovementioned two techniques even after post-hoc analysis. However, the rest of the *p-values* in Table 5.9 are not smaller than *Bonferroni_*α value.

Chapter Summary

The chapter discussed feature extraction with partial coefficients of image transforms. Five different transforms—namely, discrete cosine transform (DCT), Walsh transform, Kekre transform, discrete sine transform (DST) and Hartley transform—have been discussed and are implemented for feature extraction using partial transform coefficients for each of the techniques. The extracted feature vectors are evaluated for classification results on four different public datasets—namely, Wang, OT-Scene, Corel and Caltech. Four different classifiers—namely, K Nearest Neighbor (KNN), Ripple-Down Rule (RIDOR), Artificial Neural Network (ANN) and Support Vector Machine (SVM)—are used for assessing the classification performances. Further, the classification results with each of the feature extraction techniques are compared. The highest classification performance is shown by feature extraction with partial DCT coefficients. Hence, it is compared to the existing techniques where it has outperformed all of them. The classification results with the proposed method have also revealed statistical significance of increased classification over the state-of-the art techniques.

References

1. Das, R., Thepade, S. and Ghosh, S., 2015. Content based image recognition by information fusion with multiview features. *International Journal of Information Technology and Computer Science*, 7(10):61–73.
2. Thepade, S., Das, R. and Ghosh, S., 2017. Decision fusion-based approach for content-based image classification. *International Journal of Intelligent Computing and Cybernetics*, 10(3): 310–331.
3. Thepade, S., Das, R. and Ghosh, S., 2014. Feature extraction with ordered mean values for content based image classification. Advances in Computer Engineering, vol. 2014, Article ID 454876, 15 pages: doi:10.1155/2014/454876
4. Das, R., Thepade, S. and Ghosh, S., 2016. Framework for content-based image identification with standardized multiview features. *Etri Journal*, 38(1): 174–184.
5. Liu. C., 2013. A new finger vein feature extraction algorithm. In IEEE 6th International Congress on Image and Signal Processing (CISP), pp. 395–399.
6. Ramírez-Ortegón, M. A. and Rojas, R., 2010. Unsupervised evaluation methods based on local gray-intensity variances for binarization of historical documents. In Proceedings of the International Conference on Pattern Recognition, pp. 2029–2032.
7. Yanli Y. and Zhenxing Z., 2012. A novel local threshold binarization method for QR image. In IET International Conference on Automatic Control and Artificial Intelligence (ACAI), pp. 224–227.
8. Shaikh, S. H., Maiti, A. K. and Chaki, N., 2013. A new image binarization method using iterative partitioning. *Machine Vision and Applications*, 24(2): 337–350.

6

Content-Based Feature Extraction: Morphological Operators

6.1 Prelude

The term *morphology* refers to a distinct contour, formation or appearance. The objects to be identified in an image exhibit all the aforesaid properties. Hence, the characteristics can be well utilized to design robust descriptors for content-based image classification. Morphological operations in image processing is the collection of multiple operations based on set theory to comprehend a range of activities embracing boundary extraction, filling of small holes present in images and noise removal from images [1]. Analysis of the shape of an object present in an image can be efficiently performed by applying mathematical morphology due to its origin from set theory. Morphological elements consider a mask that is a binary image as structuring element to be provided as an input by the morphological operators. Flexible applications of all the set operators, such as intersection, union, inclusion and complement, are possible for image data. Dilation and erosion in morphology are considered to be the fundamental functions. Initially, the image is to be processed by the effect of swelling by applying pixels to the object boundaries with dilation. Further, the object shape shrinking effect is to be performed with erosion by removing the boundary pixels. The corresponding pixel in the input image is to be compared to the neighboring pixels to define the value of each pixel in the output image. This chapter introduces two different techniques of feature extraction using morphological operators—namely, top-hat transform and bottom-hat transform. The two techniques are implemented based on the block truncation coding (BTC) approach. The descriptors are generated by calculating the mean of the separated foreground and the background pixels by applying morphological operator. The size of feature vectors for each color component is 2. Thus, for three color components, the size of feature vector is $(3 \times 2) = 6$, which is obtained by applying morphological operators for feature extraction. An illustration is provided in Fig. 6.1.

Original Image

Dilated Image   Eroded Image

FIGURE 6.1
Dilation and Erosion as a Morphological Function.

6.2 Top-Hat Transform

Top-hat transform is considered a variant of grayscale opening and closing operations that is critical in generating only the bright peaks of an image [2]. It is also known as the *peak detector* and a step-wise demonstration of its execution is given as follows:

- Apply the grayscale opening operation to an image.
- Peak = Original image – opened image.
- Display the peak.
- Exit.

The top-hat transform technique is applied on each color component Red (R), Green (G) and Blue (B) of the test images of Wang, OT-Scene, Corel and Caltech datasets. A sample illustration has been given in Fig. 6.2.

Two different clusters comprising the values representing the peak and the values of the background, respectively, are formed. The mean and standard deviation for the two clusters are derived for computing the feature vectors of

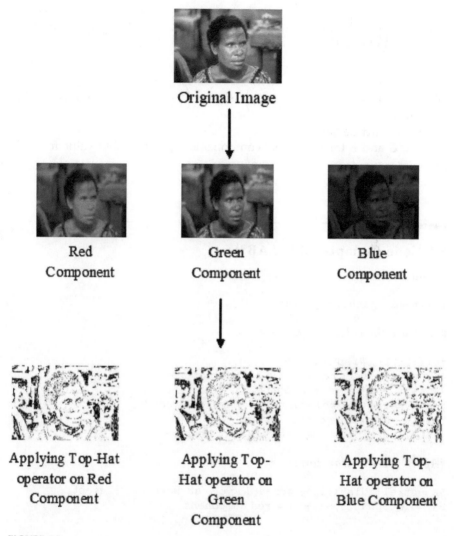

FIGURE 6.2
Application of Morphological Operator Top-Hat.

the images as in equations 6.1 and 6.2. The feature vector size is 6 for each image with 2 feature vectors for each color component.

$$x_{hiF.V.} = \left(\left(\frac{1}{m * n} \right) * \left(\sum_{I=1}^{m} \sum_{j=1}^{n} x_{hi}(i, j) \right) \right) + \left(\sqrt{\frac{1}{(m * n)} \sum_{i=1}^{m} \sum_{j=1}^{n} (x_{hi}(i, j) - \mu_{hi})^2} \right)$$

(6.1)

$$x_{loF.V.} = \left(\left(\frac{1}{(m*n)}\right) * \sum_{i=1}^{m}\sum_{j=1}^{n} x_{lo}(i, j)\right) + \left(\sqrt{\frac{1}{m*n}\sum_{i=1}^{m}\sum_{j=1}^{n}(x_{lo}(i, j) - \mu_{lo})^2}\right)$$

$$(6.2)$$

where, $\mu = \left(\frac{1}{(m*n)} * \sum_{i=1}^{m}\sum_{j=1}^{n} x(i, j)\right)$ and $\sigma = \left(\sqrt{\frac{1}{(m*n)}\sum_{i=1}^{m}\sum_{j=1}^{n}(x(i, j) - \mu)^2}\right)$

μ = mean

σ = standard deviation

x = R, G and B for individual components, T_x = threshold value for each pixel

6.3 Code Example (MATLAB®)

#Read an image

```
i=imread('path\imagefile');
```

#Separate the color components

```
r=i(:,:,1); #red component
[m n p]=size(r); #dimension of red component
g=i(:,:,2); #green component
[x y z]=size(g); #dimension of green component
b=i(:,:,3); #blue component
[j k l]=size(b); #dimension of blue component
```

#Apply top-hat transform

```
SE = strel('disk',12); #creating the structuring element
BWr = imtophat(r,SE); #r = red component
```

6.4 Coding Exercise

The previous code example applied top-hat transform on the red component of an image. Refer to the example to complete the following assignments:

- *Apply top-hat transform on Green and Blue color components.*
- *Extract features from separated background and foreground obtained after applying transform.*

TABLE 6.1

Classification Performances by Feature Extraction Using Top-Hat Transform with Four Different Public Datasets using Four Different Classifiers

Datasets	Metrics	KNN	RIDOR	ANN	SVM
Wang	Precision	0.595	0.508	0.617	0.561
	Recall	0.571	0.508	0.62	0.559
	MR	0.079	0.088	0.074	0.08
	F1-Score	0.57	0.507	0.617	0.555
OT-Scene	Precision	0.413	0.375	0.465	0.409
	Recall	0.414	0.378	0.461	0.413
	MR	0.096	0.1	0.092	0.12
	F1-Score	0.408	0.376	0.456	0.389
Corel	Precision	0.672	0.727	0.763	0.557
	Recall	0.662	0.722	0.763	0.572
	MR	0.071	0.055	0.048	0.084
	F1-Score	0.643	0.722	0.758	0.525
Caltech	Precision	0.402	0.411	0.482	0.328
	Recall	0.507	0.459	0.556	0.478
	MR	0.094	0.092	0.087	0.1
	F1-Score	0.415	0.43	0.506	0.358

[Observation: Highest classification results for all the datasets are observed with ANN classifier.]

The derived features from the four different public datasets are evaluated for classification performance by 10-fold cross-validation with four different classifiers—namely, K Nearest Neighbour (KNN), Ripple-Down Rule (RIDOR), Artificial Neural Network (ANN) and Support Vector Machine (SVM). The results are given in Table 6.1.

6.5 Bottom-Hat Transform

Bottom-hat transform is considered the dual of top-hat transform discussed in Section 6.2. A significant morphological transform termed as *opening* corresponds to *erosion* followed by *dilation* and is designated by $f \circ K$. *Dilation* followed by an *erosion* is known as *closing* and is denoted by $..$, which is the dual of *opening* [3]. *Top-hat* transform is defined as the residual of the *opening* compared to the original signal and is denoted by $f - (f \circ K)$. The bottom-hat transform is defined as the residual of a *closing* compared to the original signal f, i.e., $f - (f \bullet K)$.

The area of interest for feature extraction in each of the Red (R), Green (G) and Blue (B) color components of the images is located by means of bottom-

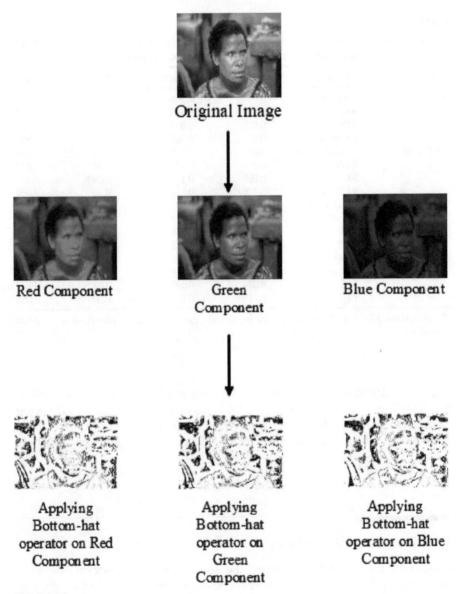

FIGURE 6.3
Application of Morphological Operator Bottom-Hat.

hat transform from four different public datasets—namely, Wang, OT-Scene, Corel and Caltech. A sample illustration is given in Fig. 6.3.

Two different clusters are formed by grouping gray values of the contour region and the background portion of individual color components of the image.

The mean and standard deviation for the two clusters are derived for computing the feature vectors of the images as in equations 6.1 and 6.2. The feature vector size is 6 for each image with 2 feature vectors for each color component.

6.6 Code Example (MATLAB)

#Read an image

```
zi=imread('path\imagefile');
```

#Separate the color components

```
r=i(:,:,1); #red component
[m n p]=size(r); #dimension of red component
g=i(:,:,2); #green component
[x y z]=size(g); #dimension of green component
b=i(:,:,3); #blue component
[j k l]=size(b); #dimension of blue component
```

#Applying bot-hat transform

```
SE = strel('disk',12); #creating the structuring element
BWr = imbothat(r,SE); #r = red component
```

6.7 Coding Exercise

The previous code example applies bottom-hat transform on the red component of an image. Refer to the example to complete the following assignments:

- *Apply bot-hat transform on Green and Blue color components.*
- *Extract features from separated background and foreground obtained after applying transform.*

Hereafter, calculation of average values for diverse metrics is carried out to observe Precision, Recall, Misclassification Rate (MR) and F1-Score for classification with four different classifiers—namely, KNN, RIDOR, ANN and SVM—with feature extraction applying bottom-hat transform from four different public datasets.

The comparative results are given in Table 6.2.

TABLE 6.2

Classification Performances by Feature Extraction Using Bottom-Hat Transform with Four Different Public Datasets using Four Different Classifiers

Datasets	Metrics	KNN	RIDOR	ANN	SVM
Wang	Precision	0.767	0.726	0.826	0.72
	Recall	0.761	0.729	0.83	0.714
	MR	0.052	0.059	0.043	0.066
	F1-Score	0.761	0.725	0.826	0.707
OT-Scene	Precision	0.602	0.615	0.65	0.481
	Recall	0.589	0.613	0.637	0.498
	MR	0.079	0.078	0.074	0.093
	F1-Score	0.589	0.613	0.635	0.451
Corel	Precision	0.691	0.751	0.792	0.594
	Recall	0.671	0.76	0.796	0.604
	MR	0.07	0.053	0.046	0.079
	F1-Score	0.659	0.754	0.79	0.57
Caltech	Precision	0.514	0.533	0.64	0.349
	Recall	0.573	0.566	0.665	0.537
	MR	0.089	0.08	0.076	0.095
	F1-Score	0.498	0.546	0.65	0.417

[**Observation:** Highest classification results for all the datasets are observed with ANN classifier.]

6.8 Comparison of Proposed Techniques

This chapter discusses two different techniques of morphological feature extraction—namely, top-hat transform and bottom-hat transform. The techniques are tested for classification results on four different public da-tasets using four different classifiers and the results are given in Tables 6.1 and 6.2. The two techniques are compared in Figs. 6.4–6.7 based on Precision, Recall, MR and F1-Score respectively for all the four datasets and classifiers.

The comparison shown in Figs. 6.4–6.7 has clearly revealed that classification by feature extraction with bottom-hat transform has better Precision, Recall and F1-Score values and lesser MR compared to top-hat transform.

Feature extraction using bottom-hat transform is carried out exclusively from the extracted area of interest in an image. This has led to robust feature extraction because the structuring element in bottom-hat transform removes the image portions that do not fit to it during both opening and closing operations. On the contrary, the top-hat transform is used for light object with dark background and the removed image parts comprise portions smaller than structuring elements. Thus, the region of interest for the given

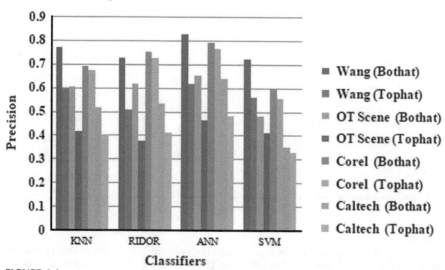

FIGURE 6.4
Comparison of Precision Values of Classification with Top-Hat transform and Bottom-Hat Transform Method of Feature Extraction for All Four Datasets using Four Different Classifiers. [**Observation:** Bottom-hat transform has exceeded top-hat transform in all four datasets.]

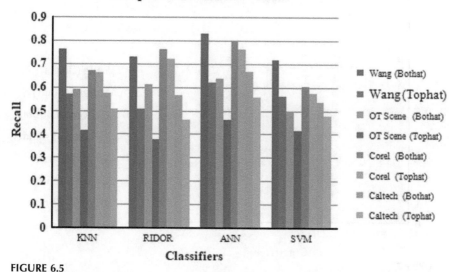

FIGURE 6.5
Comparison of Recall Values of Classification with Top-Hat Transform and Bottom-Hat Transform Method of Feature Extraction for All Four Datasets using Four Different Classifiers. [**Observation:** Bottom-hat transform has exceeded top-hat transform in all four datasets.]

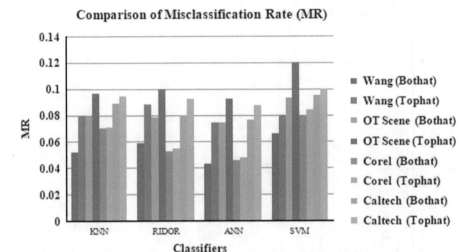

FIGURE 6.6
Comparison of MR of Classification with Top-Hat Transform and Bottom-Hat Transform
Method of Feature Extraction for All Four Datasets using Four Different Classifiers.
[**Observation:** Bottom-hat transform has exceeded top-hat transform in all four datasets.]

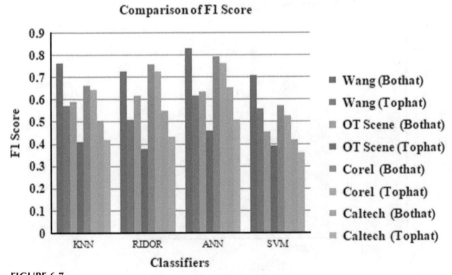

FIGURE 6.7
Comparison of F1-Score of Classification with Top-Hat Transform and Bottom-Hat Transform
Method of Feature Extraction for All Four Datasets using Four Different Classifiers.
[**Observation:** Bottom-hat transform has exceeded top-hat transform in all four datasets.]

types of images in this work is not properly identified for feature extraction by top-hat transform. Hence, classification results for feature extraction with bottom-hat transform outperformed the top-hat transform.

The comparison in Fig. 6.7 has shown better Precision, Recall, F1-Score values and lesser MR for classification by feature extraction with bottom-hat transform with respect to the existing techniques.

6.9 Comparison with Existing Methods

Feature extraction using bottom-hat transform is superior to the top-hat transform in Section 6.3. Hence, it is compared for classification results with respect to the state-of-the art techniques [4–7] in Fig. 6.8. The comparison is done with the Wang dataset and KNN classifier as the existing techniques is evaluated using the same environment.

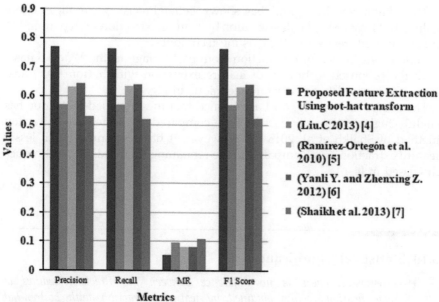

FIGURE 6.8
Comparison of Classification Results by Feature Extraction Using Bottom-Hat Technique to the Existing Techniques.

[Observation: Classification results for feature extraction with bottom-hat transform has exceeded the classification results of the existing techniques.]

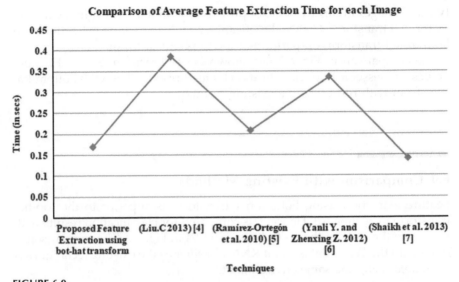

FIGURE 6.9
Comparison of Average Time Consumed for Feature Extraction.
[**Observation:** Feature extraction with the proposed technique of feature extraction with bottom-hat transform has consumed higher average time compared to Shaikh et al. 2013.]

The comparison in Fig. 6.7 has shown better Precision, Recall, F1-Score values and lesser MR for classification by feature extraction with bottom-hat transform with respect to the existing techniques.

Time taken for feature extraction from each image in the Wang dataset with the proposed technique of feature extraction with bottom-hat transform is compared to the existing technique in Fig. 6.9.

The comparison in Fig. 6.9 has revealed that the proposed technique has slightly higher feature extraction time compared to the feature extraction method of Shaikh et al. 2013 [7]. However, it has consumed much lesser feature extraction time compared to the remaining three existing techniques discussed in [4–6].

6.10 Statistical Significance

HYPOTHESIS 1: *There is no difference between the Precision values of classification results obtained by feature extraction with bottom-hat transform with respect to the existing techniques.*

Difference in precision values of classification results by feature extraction with bottom-hat transform and the existing techniques is evaluated with a

TABLE 6.3

Value of Mean Standard Deviation and Standard Error of the Mean for Precision Values of Classification by Feature Extraction with Bottom-Hat Transform

Proposed Technique	Mean	SD	SEM
Feature extraction with bottom-hat transform	0.753	0.117	0.04

TABLE 6.4

t-test for Evaluating the Significance of Feature Extraction using Bottom-Hat Transform

Comparison	Mean	SD	SEM	*t-calc*	*p*-value	Significance
Liu, 2013	*0.626*	*0.22*	*0.073*	*2.03*	*0.077*	*Insignificant*
Ramírez-Ortegón et al., 2010	*0.601*	*0.24*	*0.08*	*2.2*	*0.058*	*Insignificant*
Yanli and Zhenxing, 2012	*0.638*	*0.223*	*0.074*	*1.89*	*0.09*	*Insignificant*
Shaikh et al., 2013	*0.595*	*0.214*	*0.071*	*2.6*	*0.031*	*Significant*

[Observation: The *p* values have indicated significant differences in Precision results only for classification by feature extraction using Shaikh et al. (2013) technique of feature extraction compared to feature extraction using bottom-hat transform.]

paired *t*-test. The test is conducted to determine whether the difference in Precision values for classification is generated from a population with zero mean as the results are displayed in Table 6.4. Table 6.3 displays the mean, standard deviation (SD), and standard error of the mean (SEM) of the Precision values of classification of different categories for feature extraction using bottom-hat transform.

Calculation of *t*-calc is performed by taking the difference between two sample means considered for each comparison in Table 6.4. The strength of evidence against the null hypothesis is measured by the *p*-value.

The calculated *p* values in Table 6.4 indicate an insignificant difference in Precision results for classification with proposed method of feature extraction compared to the existing techniques, except feature extraction with Shaikh et al. (2013). Therefore, the null hypothesis is accepted for all the comparisons except for the last one in Table 6.4, where it is rejected.

However, the values of mean, SD and SEM of the existing techniques given in Table 6.4 are further compared to that of the proposed method in Table 6.3. It is observed that the proposed method has a higher mean of precision values with less SD and SEM compared to all the existing techniques. Hence, it can be concluded that the proposed method has not only shown higher Precision values, but also has consistent performance because of lesser SD and SEM compared to the existing techniques.

As a result, statistically, the supremacy of the proposed technique is established only over feature extraction with Otsu's threshold selection for content-based image classification.

The *t-test* is performed with four different comparisons; a *post-hoc* analysis is necessary to accept the significance of the derived results using the Bonferroni correction. The value of α is considered to be *0.05* for the *t-test*. The following expression calculates the Bonferroni correction value:

$$Bonferroni\_\alpha = \frac{\alpha}{Number\_of\_tests}$$

where $\alpha = 0.05$

   $Number\_of\_tests = 4$

   Therefore,

   $Bonferroni\_\alpha = 0.013$

Now, the *p-values* in Table 6.4 are compared to the newly calculated *Bonferroni_α* value of 0.013.

The comparison shows that none of the *p-values* in Table 6.4 is smaller than *Bonferroni_α* value.

Chapter Summary

This chapter discussed two different morphological operators—namely, top-hat transform and bottom-hat transform. The techniques are modified and applied for feature extraction for content-based image classification. The performance is tested with four different datasets and four different classifiers. The bottom-hat technique outclassed the top-hat technique of feature extraction based on classification results. Comparison with state-of-the-art techniques proved the bottom-hat transform is better. The statistical comparison also revealed partial superiority of the bottom-hat transform with respect to the existing techniques.

References

1. Sridhar, S., 2011. *Image Features Representation and Description, Digital Image Processing*, India Oxford University Press, New Delhi, pp. 483–486.
2. Thepade, S., Das, R. and Ghosh, S., 2017. Decision fusion-based approach for content-based image classification. *International Journal of Intelligent Computing and Cybernetics*, 10 (3): 310–331.

3. Das, R., Thepade, S. and Ghosh, S., 2016. Framework for content-based image identification with standardized multiview features. *Etri Journal*, 38 (1): 174–184.

4. Liu, C., 2013. A new finger vein feature extraction algorithm, In IEEE 6th International Congress on Image and Signal Processing (CISP), pp. 395–399.

5. Ramírez-Ortegón, M. A. and Rojas, R., 2010. Unsupervised evaluation methods based on local gray-intensity variances for binarization of historical documents. In Proceedings of the International Conference on Pattern Recognition, pp. 2029–2032.

6. Yanli, Y. and Zhenxing, Z., 2012. A novel local threshold binarization method for QR image. In IET International Conference on Automatic Control and Artificial Intelligence (ACAI), pp. 224–227.

7. Shaikh, S. H., Maiti, A. K. and Chaki, N., 2013, A new image binarization method using iterative partitioning. *Machine Vision and Applications*, 24 (2): 337–350.

References

7

Content-Based Feature Extraction: Texture Components

7.1 Prelude

A frequent property observed in an image in the form of repeating structures having some randomness is called *texture*. The elementary component of a repeating pattern is designated as basic primitive or *texel*. Placement rules or uniformity are used by the basic primitives to form repetitive patterns. In some images, the defining characteristics of regions can be important for proper analysis. Feature extraction can be done from the spatial arrangements of colors or intensities with the help of texture analysis. Some histogram distributions can have different texture representations, which can act as a tool for extraction of distinct features. Thus, the presence of texture appears in the form of repeated patterns as shown in Fig. 7.1.

Variations in intensity and color have led to the formation of textures. Textures correspond to qualitative terms, namely, coarseness, homogeneity and smoothness at the physical level. Textures are divided into micro textures and macro textures based on the size of the primitives. Macro texture or coarseness of an object is defined by large elements of primitive, whereas micro texture is formed by smaller textures. The orientation of texture elements is known as dimensionality. Two methods of feature extraction using texture content of an image are introduced in this chapter, namely, feature extraction by vector quantization (VQ) codebook representation using the Linde-Buzo-Grey (LBG) algorithm and feature extraction by the gray level co-occurrence matrix (GLCM). These techniques have enhanced the content-based image classification rate. In our approach, we have generated features by calculating code vectors from the generated clusters using the LBG algorithm. The size of the codebook thus generated is much smaller than the image dimension and is used as a feature vector for content-based image classification. In the case of GLCM, the mean and standard deviation of all the parameters, namely, energy, contrast, entropy and correlation, are considered to compute the feature vector for each image. This generates a feature

FIGURE 7.1
Sample Texture Pattern.
Source: https://upload.wikimedia.org/wikipedia/commons/
a/a9/ArtificalTexture.png (Under Creative Commons license).

size of 8 for each image which is independent of the image dimension and
has a considerably smaller footprint.

7.2 Feature Extraction by Vector Quantization Codebook Representation Using Linde-Buzo-Grey (LBG) Algorithm

The statistical color distribution of an image has been described by the vector
quantization (VQ) codebook [1]. The technique is applied on four different
public datasets for feature extraction from the test images. The datasets are
Wang dataset, OT-Scene dataset, Corel dataset and Caltech dataset. Each of
the color components, namely, red (R), green (G) and blue (B), are extracted
from the test images for applying vector quantization to generate codebook as
feature vectors. A k dimensional Euclidian space is mapped by means of
vector quantization into a finite subset. Representation of the codebook is
done by the finite set CB as in equation 7.1.

$$CB = \{Ci/i = 1,\ 2,\ \ldots.,\ N\} \tag{7.1}$$

where $Ci = (ci1, ci2, \ldots., cik)$ is a code vector and N is the size of the codebook.
The Linde-Buzo-Gray (LBG) algorithm is followed for generation of code
vectors. Primarily, the images are divided into nonoverlapping blocks and
are converted to training vector $Xi = (xi1, xi2, \ldots, xik)$. A dimension of 12 is
fixed for each training vector in the training set. The training vector is comprised of red, green and blue components of 2×2 neighboring pixels.
Henceforth, the centroid of the entire training set is calculated as the first code
vector. In addition, two trial code vectors $v1$ and $v2$ are generated by adding
and subtracting the constant error to the centroid. The nearness of each
training vector to the trial vectors is computed, and two clusters are formed
based on proximity of the training vectors to $v1$ and $v2$ as shown in Fig. 7.2.
This has created a codebook of size 2 by calculating two centroids from the
two newly formed clusters to produce two code vectors. The aforesaid process

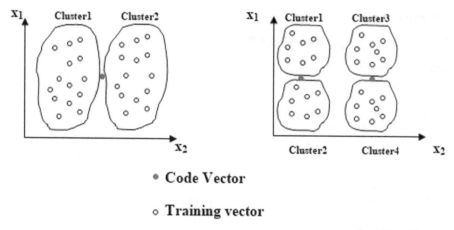

• Code Vector

○ Training vector

FIGURE 7.2
Implementation of Linde-Buzo-Grey (LBG) Algorithm.

is repeated with the centroids to generate the desired size of the codebook, which is 16 in this case.

The extracted features from the four public datasets are tested with four different classifiers, namely, K Nearest Neighbor (KNN), Ripple-Down Rule (RIDOR), Artificial Neural Network (ANN) and Support Vector Machine (SVM). The results are given in Table 7.1.

TABLE 7.1

Classification Performances by Feature Extraction using Vector Quantization with Four Different Public Datasets using Four Different Classifiers

Datasets	Metrics	KNN	RIDOR	ANN	SVM
Wang	Precision	0.900	0.900	0.928	0.888
	Recall	0.890	0.900	0.925	0.869
	MR	0.042	0.039	0.038	0.043
	F1-Score	0.884	0.900	0.925	0.862
OT-Scene	Precision	0.923	0.922	0.937	0.860
	Recall	0.919	0.922	0.936	0.860
	MR	0.038	0.039	0.036	0.044
	F1-Score	0.923	0.922	0.937	0.856
Corel	Precision	0.703	0.713	0.820	0.545
	Recall	0.658	0.668	0.819	0.499
	MR	0.057	0.073	0.046	0.091
	F1-Score	0.700	0.660	0.819	0.468
Caltech	Precision	0.725	0.781	0.863	0.714
	Recall	0.725	0.777	0.864	0.712
	MR	0.057	0.053	0.042	0.066
	F1-Score	0.707	0.778	0.862	0.714

[Observation: All four datasets have maximum classification results with the ANN classifier.]

7.3 Code Example (MATLAB®)

#Read an Image
```
i=imread('path\imagefile');
```

#Separate the Color Components
```
r=i(:,:,1); #red component
[m n p]=size(r); #dimension of red component
g=i(:,:,2); #green component
[x y z]=size(g); #dimension of green component
b=i(:,:,3); #blue component
[j k l]=size(b); #dimension of blue component
```

#Computing Trial Code Vector
```
Thr=mean2(r);
meanr1=Thr+1; #Trial code vector 1 for red component
meanr2=Thr-1; #Trial code vector 2 for red component
Thg=mean2(g);
meang1=Thg+1; #Trial code vector 1 for green component
meang2=Thg-1; #Trial code vector 2 for green component
Thb=mean2(b);
meanb1=Thb+1; #Trial code vector 1 for blue component
meanb2=Thb-1; #Trial code vector 2 for blue component
```

#Sample Code for Codebook Representation from Red Component
```
for a1=1:m
for a2=1:n
sub1=abs(meanr1-i1(a1,a2));
sub2=abs(meanr2-i1(a1,a2));
if sub1>sub2
c1=c1+1;
arr1(c1,1)=i1(a1,a2);
end
if sub1<=sub2
c2=c2+1;
arr2(c2,1)=i1(a1,a2);
end
end
end
```

7.4 Coding Exercise

The above code example illustrates the process of codebook generation using the LBG algorithm. Refer to the example and try to complete the following assignment:

- *Apply the LBG algorithm on red, green and blue color components for codebook generation.*

7.5 Feature Extraction by Gray Level Co-occurrence Matrix (GLCM)

Texture features have the characteristics to describe the visual pattern based on spatial definition of images. Identifying the specific texture in an image is carried out by representing the texture as a two-dimensional gray level variation known as the gray level co-occurrence matrix (GLCM) [1]. The GLCM is defined as a statistical method for discovering the textures that consider the spatial relationships of pixels. The GLCM function represents the texture of an image by computing the frequency of existence of pixel pairs with specific values and with specific pixel re-lationships in an image followed by extraction of statistical measures from the matrix. (7.2) defines the normalized probability of the co-occurrence matrix.

$$P_\delta(i, j) = \frac{\#\{\mid (x, y), \ (x + d, \ y + d) \mid \ \in S \mid f(x, y) = i, f(x + d, \ y + d) = j\}}{\#S}$$

$$(7.2)$$

where $x,y = 0, 1, ..., N - 1$ are pixel coordinates, $i,j = 0, 1, ..., L - 1$ are gray levels, S = pixel pairs in image with certain relationship, $\#S$ = number of elements in S and $P_\delta(i,j)$ = probability density.

Different directions are considered for computation of GLCM, such as $\delta = 0°$, $\delta = 45°$, $\delta = 90°$ and $\delta = 135°$. Four statistical parameters for energy, contrast, entropy and correlation are calculated based on the GLCM as shown in equations 7.3–7.6. The repetition of pixel pairs is measured by the energy or angular second moment (ASM). The variance of gray level is quantified by contrast. The disorder of the image is measured by entropy, and correlation has returned a measure of how correlated a pixel is to its neighbor over the whole image.

$$ASM = \sum\sum P^2(i, j) \qquad\qquad (7.3)$$

$$CON = \sum\sum (i - j)^2 P(i, j) \qquad (7.4)$$

$$ENT = -\sum \sum P(i, j) \log [P(i, j)] \qquad (7.5)$$

$$COR = \frac{\sum\sum ijP(i - j) - \mu_x\mu_y}{\sigma_x\sigma_y} \qquad (7.6)$$

where μ_x and μ_y and σ_x and σ_y denote the mean and standard deviation for P_X and P_Y, respectively; P_X = sum of each row of co-occurrence matrix and P_Y = sum of each column of co-occurrence matrix. The mean and standard deviation of all the parameters are used to compute the feature vector for each image and the dimension of feature vector is 8 per image.

The technique of feature extraction is applied on four public datasets, namely, Wang dataset, OT-Scene dataset, Corel dataset and Caltech dataset, and is tested for classification results with four different classifiers namely, K Nearest Neighbor (KNN), Ripple-Down Rule (RIDOR), Artificial Neural Network (ANN) and Support Vector Machine (SVM). The results are given in Table 7.2.

TABLE 7.2

Classification Performances by Feature Extraction using GLCM with Four Different Public Datasets using Four Different Classifiers

Datasets	Metrics	KNN	RIDOR	ANN	SVM
Wang	Precision	0.615	0.481	0.633	0.565
	Recall	0.617	0.492	0.634	0.566
	MR	0.079	0.092	0.076	0.080
	F1-Score	0.604	0.485	0.631	0.553
OT-Scene	Precision	0.528	0.461	0.578	0.475
	Recall	0.526	0.462	0.580	0.483
	MR	0.082	0.093	0.080	0.094
	F1-Score	0.526	0.461	0.573	0.448
Corel	Precision	0.535	0.444	0.464	0.468
	Recall	0.548	0.441	0.456	0.475
	MR	0.082	0.094	0.097	0.093
	F1-Score	0.524	0.442	0.400	0.449
Caltech	Precision	0.436	0.577	0.527	0.369
	Recall	0.544	0.587	0.583	0.506
	MR	0.090	0.080	0.079	0.100
	F1-Score	0.466	0.578	0.544	0.388

[Observation: The Wang and OT-Scene datasets show maximum classification performance with the ANN classifier, the Corel dataset shows maximum classification with the KNN classifier and the Caltech dataset shows maximum classification with the RIDOR classifier.]

7.6 Code Example (MATLAB)

#Read an Image

```
i=imread('path\imagefile');
```

#Separate the Color Components

```
r=i(:,:,1); #red component
[m n p]=size(r); #dimension of red component
g=i(:,:,2); #green component
[x y z]=size(g); #dimension of green component
b=i(:,:,3); #blue component
[j k l]=size(b); #dimension of blue component
```

GLCM Features with 'Offset', [0 1] for Red Color Component

```
glcm1 = graycomatrix(r,'Offset',[0 1]);
stats1 = graycoprops(glcm1,{'Contrast','Homogeneity','Energy',
'Correlation'});
value1 = getfield(stats1, 'Contrast');
value2 = getfield(stats1, 'Homogeneity');
value3 = getfield(stats1, 'Energy');
value4 = getfield(stats1, 'Correlation');
```

GLCM Features with 'Offset', [-1 1] for Red Color Component

```
glcm4 = graycomatrix(r,'Offset',[-1 1]);
stats4 = graycoprops(glcm4,{'Contrast','Homogeneity','Energy',
'Correlation'});
value13 = getfield(stats4, 'Contrast');
value14 = getfield(stats4, 'Homogeneity');
value15 = getfield(stats4, 'Energy');
value16 = getfield(stats4, 'Correlation');
```

7.7 Coding Exercise

The above code example illustrates the process of feature extraction with GLCM using two different offsets for the red component. Refer to the example and try to complete the following assignments:

- *Apply GLCM on the red component to extract features for all offset values.*
- *Apply GLCM on green and blue components to extract features using all offset values.*

7.8 Evaluation of Proposed Techniques

Two techniques of texture content-based feature extractions were introduced in the previous subsections. The techniques were tested for classification results with four different widely used public datasets. Four different classifiers were used for the comparison. The two techniques are compared based on Precision, Recall, Misclassification Rate (MR) and F1-Score values in Figs. 7.3–7.6.

The comparisons based on Precision, Recall, MR and F1-Score shown in Figs. 7.3–7.6 establish feature extraction with VQ using LBG as a superior technique compared to that of the GLCM method of feature extraction for content-based image classification.

The LBG technique divides the gray values into separate clusters based on likelihood and separates the given test image into local partitions for feature extraction. This has locally addressed the adversities of test images, namely, variance in gray levels within the object and the background, inadequate contrast, etc., while calculating the code vectors from each cluster. This has further facilitated robust feature extraction from the region of interest in the test images.

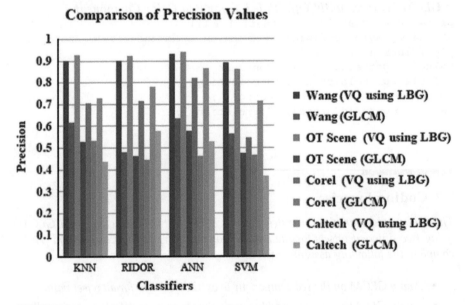

FIGURE 7.3

Comparison of Precision Values of Classification by Feature Extraction with VQ using LBG and GLCM for All Four Datasets using Four Different Classifiers.

[**Observation:** Classification by feature extraction with VQ using LBG has higher Precision values compared to GLCM.]

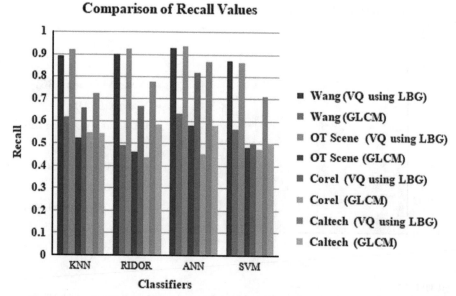

FIGURE 7.4
Comparison of Recall Values of Classification by Feature Extraction with VQ using LBG and GLCM for All Four Datasets using Four Different Classifiers.
[**Observation:** Classification by feature extraction with VQ using LBG has higher Recall values compared to GLCM.]

A conventional approach has been adopted in case of GLCM for feature extraction in the current method. The technique uses symmetrical co-occurrence matrices. However, use of asymmetrical co-occurrence matrices can improve identification of resourceful features in the test images considered, since they contain texture orientation information in opposite direction. Moreover, concatenation of two-dimensional matrices is desirable over summing them up into a single matrix for better understanding of feature distribution.

7.9 Comparison with Existing Methods

Feature extraction with VQ using LBG has been established as a superior technique for content-based classification compared to GLCM as discussed in Section 7.3. Hence, the technique is compared to the state-of-the-art techniques of feature extraction [2–5] for classification results based on Precision, Recall, MR and F1-Score as shown in Fig. 7.7.

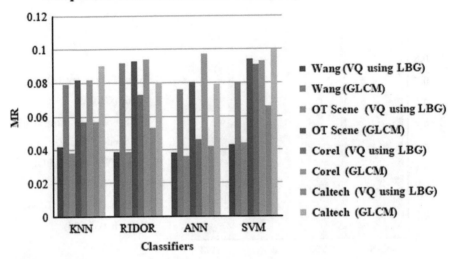

FIGURE 7.5
Comparison of Misclassification Rate (MR) of Classification by Feature Extraction with VQ using LBG and GLCM for All Four Datasets using Four Different Classifiers.
[**Observation:** Classification by feature extraction with VQ using LBG has lower MR values compared to GLCM.]

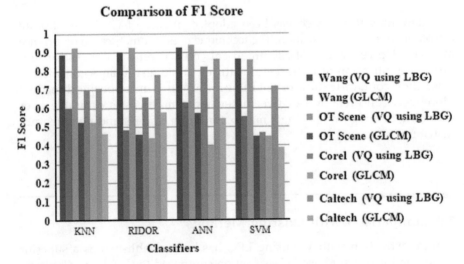

FIGURE 7.6
Comparison of F1 Scores of Classification by Feature Extraction with VQ using LBG and GLCM for All Four Datasets using Four Different Classifiers.
[**Observation:** Classification by feature extraction with VQ using LBG has higher F1-Score values compared to GLCM.]

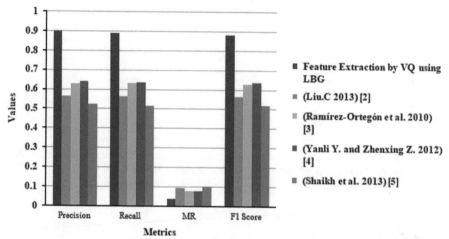

Comparison for Classification results by feature extraction using vector quantization (VQ) using LBG technique with respect to state-of-the art techniques of feature extraction

FIGURE 7.7
Comparison of Classification Results by Feature Extraction using Bottom-Hat Technique to That of Existing Techniques.
[**Observation:** Classification results for feature extraction with bottom-hat transform exceed the classification results for the existing techniques.]

The results for classification by feature extraction with VQ using LBG in Fig. 7.7 have clearly outclassed the existing techniques, profoundly enhanced the rate of Precision, Recall and F1-Score and reduced the MR for content-based image classification.

Finally, the average time taken for feature extraction from each image in a Wang dataset with VQ using the LBG technique is compared to the existing techniques discussed in References [2–5]. A comparison is shown in Fig. 7.8.

The comparison in Fig. 7.8 shows that feature extraction with VQ using the LBG technique consumes maximum average time compared to the existing techniques discussed in References [2–5].

7.10 Statistical Significance

HYPOTHESIS 7.1: *There is no difference between the Precision values of classification results obtained by feature extraction with VQ using LBG with respect to the existing techniques.*

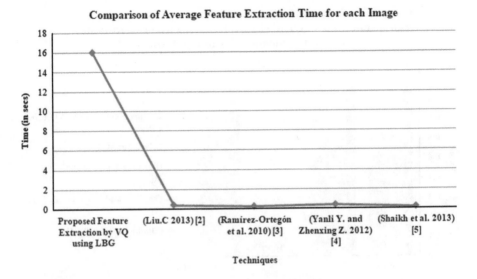

FIGURE 7.8
Comparison of Average Time Consumed for Feature Extraction.
[Observation: Feature extraction with VQ using the LBG technique consumes maximum average time compared to existing techniques.]

A paired t-test is carried out to establish the difference in Precision values for classification by feature extraction with VQ using LBG compared to the existing techniques. Initially, the mean, standard deviation (SD) and standard error of the mean (SEM) of the Precision values of classification of different categories for feature extraction with VQ using LBG are calculated in Table 7.3.

The results for the t-test to determine whether the difference in Precision values for classification is generated from a population with zero mean are displayed in Table 7.4.

The p-values in Table 7.4 indicate significant difference in Precision results for classification by feature extraction with VQ using LBG compared to the existing feature extraction techniques. Comparison of the mean, SD and SEM of Tables 7.3 and 7.4 shows that the mean value of Precision of

TABLE 7.3

Value of Mean, Standard Deviation and Standard Error of the Mean for Precision Values of Classification by Feature Extraction with VQ using LBG

Proposed Technique	Mean	Standard Deviation (SD)	Standard Error of the Mean (SEM)
Feature extraction with VQ using LBG	0.899	0.085	0.020

TABLE 7.4

t-test for Evaluating Significance of Feature Extraction with VQ using LBG

Existing Techniques	Mean	Standard Deviation (SD)	Standard Error of the Mean (SEM)	*t*-calc	*p*-Value	Significance
Liu, 2013	0.626	0.220	0.073	3.550	0.007	Significant
Ramírez-Ortegón et al., 2010	0.601	0.240	0.080	3.770	0.005	Significant
Yanli and Zhenxing, 2012	0.638	0.223	0.074	3.190	0.012	Significant
Shaikh et al., 2013	0.595	0.214	0.071	3.870	0.004	Significant

[Observation: The *p*-values indicate significant difference in Precision results for classification with all the existing feature extraction techniques compared to the feature extraction with VQ using LBG.]

the proposed technique is better than the existing techniques. On the other hand, the SD and SEM have a lesser value for the proposed technique compared to the existing techniques. Therefore, the proposed technique can be considered to have better consistency compared to the existing techniques. Hence, classification by feature extraction with VQ using LBG has outclassed the state-of-the-art techniques in classification performance for content-based image classification.

The *t*-test is performed with four different comparisons; a post hoc analysis must be conducted to accept the significance of the derived results using the Bonferroni correction. The value of α is considered to be 0.05 for the *t*-test. The expression to calculate the Bonferroni correction value is given as follows:

$$Bonferroni\_\alpha = \frac{\alpha}{Number\_of\_tests}$$

where $\alpha = 0.05$ and $Number\_of\_tests = 4$. Therefore,

$Bonferroni\_\alpha = 0.013$

Now, the *p*-values in Table 7.4 are compared to the newly calculated .. value of 0.013.

The comparison shows that all the *p*-values in Table 7.4 are smaller than the *Bonferroni_α* value. Thus, the proposed technique has significantly different classification results even after post hoc analysis compared to all the existing techniques in Table 7.4.

Chapter Summary

This chapter presented two feature extraction techniques, namely, VQ using LBG and GLCM for content-based image classification. The techniques were tested with four different public datasets using four different classifiers. Performance comparison of the two techniques has established that the former is better than the latter. Classification performance by feature extraction with VQ using LBG was compared to state-of-the-art techniques of feature extraction for classification results. The introduced technique surpassed the performance of the existing techniques and revealed statistical significance of improved content-based image classification.

References

1. Das, R., Thepade, S., and Ghosh, S., 2017. Decision Fusion for Classification of Content Based Image Data. In *Transactions on Computational Science XXIX*, Springer, Berlin, Heidelberg, pp. 121–138.
2. Liu. C., 2013. A new finger vein feature extraction algorithm. In IEEE 6th International Congress on Image and Signal Processing (CISP), pp. 395–399.
3. Ramírez-Ortegón, M. A. and Rojas, R., 2010. Unsupervised evaluation methods based on local gray-intensity variances for binarization of historical documents. In Proceedings -International Conference on Pattern Recognition, pp. 2029–2032.
4. Yanli, Y. and Zhenxing, Z., 2012. A novel local threshold binarization method for QR image. In IET International Conference on Automatic Control and Artificial Intelligence (ACAI), pp. 224–227.
5. Shaikh, S. H., Maiti, A. K. and Chaki, N., 2013. A new image binarization method using iterative partitioning. *Machine Vision and Applications*, 24 (2): 337–350.

8

Fusion-Based Classification: A Comparison of Early Fusion and Late Fusion Architecture for Content-Based Features

8.1 Prelude

Preceding years of studies on content-based image classification have emphasized the extraction of a single feature to represent image content. Achieving satisfactory classification results with a distinct image feature is challenging since other important features are ignored during the recognition task. An image comprises a rich feature set that is diverse in nature and can jointly create a better representation of the image content. Consequently, combining different features enhances accuracy in classification results.

Recent literature has applied the fusion framework by combining color and texture features. All the low-level features have been combined in content-based color image identification. The proposed technique has acquired percentages of a small number of dominant colors in the image by using color quantization algorithm with clusters. Shape descriptor and texture features have been extracted by using the pseudo-Zernike moments and steerable filter decomposition. Elevated accuracy in image identification has been achieved through merging of color, texture and shape. Three image features of color histogram, gray level co-occurrence matrix and Zernike Moments have been fused in the approach. Merging of low-level features has led to improved accuracy. Structure elements descriptors have been proposed to extract color and texture of images as features. Amalgamation of statistical and structural texture description has been carried out to represent spatial correlation of color and texture.

This chapter discusses two different approaches for the fusion-based image classification. Different descriptors are used to represent multiple features of a single image useful for improved content-based classification [1,2]. The framework for classification is carefully crafted to ensure the fusion of a classification decision using z score normalization. Each of these approaches is described in the following sections.

Four different datasets, namely, Wang dataset, OT-Scene dataset, Corel dataset and Caltech dataset, have been used for the purpose of experimentation. Primarily, two different feature extraction techniques are implemented to initiate the process of content-based image identification. The features are extracted by binarizing the images using Sauvola's local threshold selection method and by applying discrete cosine transform to generate partial transform coefficients of the images as feature vectors. The following subsections detail each feature extraction technique including preprocessing of image data.

8.2 Image Preprocessing

Augmented image data have the ability to provide robust feature vectors. The images are processed to create odd varieties by following equation 8.1. The resultant image is shown in Fig. 8.1.

$$\text{Im}_{odd} = \frac{(\text{Im}-\text{I}\tilde{\text{m}})}{2} \tag{8.1}$$

The contribution of each bit in the configuration of the odd image is reviewed by means of bit plane slicing. Extraction of 8 bit planes from the set of generic and odd image variety has been performed to represent the intensity value of each pixel by an 8 bit binary vector as shown in Fig. 8.2. The expression for bit plane slicing is given in equation 8.2.

$$I_{bitplane}(i, j) = \text{Re } m\left\{\frac{1}{2}floor\left[\frac{1}{2^i}I(i, j)\right]\right\} \tag{8.2}$$

The fused image is denoted by $I(i,j)$, the remainder is symbolized by Rem and $floor(I)$ stands for rounding the elements to the I nearest integers less than or equal to I.

A binary matrix is embodied by each bit plane that has contributed to the formation of image slices for subsequent bit planes as shown in Fig. 8.3.

Generic Image   Flipped Image   Generic - Flipped (Odd)

FIGURE 8.1
Odd Image Creation.

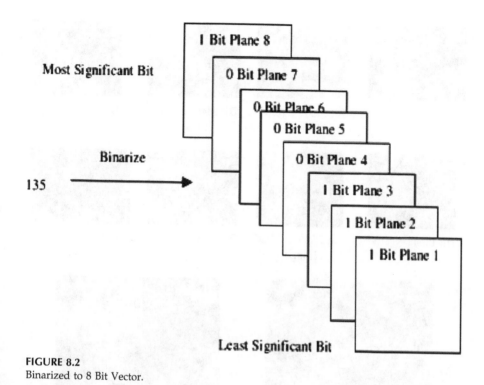

FIGURE 8.2
Binarized to 8 Bit Vector.

8.3 Feature Extraction with Image Binarization

Images are three-dimensional (3-D) in nature and have three basic color components, namely, red (R), green (G) and blue (B). The algorithm designed for feature extraction expects two-dimensional (2-D) images, which has been made possible by differentiating the red, green and blue color channels of the image. The color components are derived from the odd image variety produced with bit planes 5, 6, 7 and 8 as shown in Fig. 8.3.

A popular threshold selection technique, called Sauvola's local threshold selection technique, has been implemented for the purpose of binarization. The technique is an enhancement of Niblack's local threshold selection algorithm and is based on local-variance. Feature extraction using local threshold selection is preferred over the global threshold selection method for acquiring minute details of the image for better classification results. Equation 8.3 used for threshold selection is as follows:

where μ = mean, σ = standard deviation, $k = 0.5$ and $R = 128$.

$$T(i, j) = \mu(i, j)^* \left[1 + k \left(\frac{\sigma(i, j)}{R} - 1 \right) \right] \tag{8.3}$$

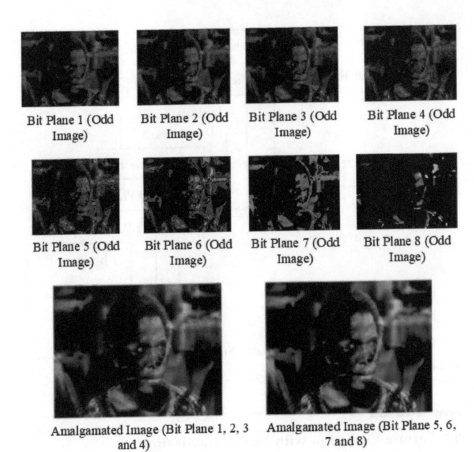

Bit Plane 1 (Odd Image)   Bit Plane 2 (Odd Image)   Bit Plane 3 (Odd Image)   Bit Plane 4 (Odd Image)

Bit Plane 5 (Odd Image)   Bit Plane 6 (Odd Image)   Bit Plane 7 (Odd Image)   Bit Plane 8 (Odd Image)

Amalgamated Image (Bit Plane 1, 2, 3 and 4)   Amalgamated Image (Bit Plane 5, 6, 7 and 8)

FIGURE 8.3
Process of Bit Plane Slicing and Amalgamation for Odd Image.

Binarization using Sauvola's local threshold selection method results in the images shown in Fig. 8.4.

The threshold value is calculated for a specified window where each pixel is compared for values higher than the threshold and lower than the threshold. The pixel having a higher value compared to the threshold is assigned a value of 1; otherwise, the value assigned is 0. The operation is shown in equation 8.4 as follows:

$$Bitmapx\,(i,\,j) = \begin{cases} 1, & if\ x(i,j) > T_x \\ 0, & if\ x(i,j) < =T_x \end{cases} \qquad (8.4)$$

where $x = R$, G and B and T_x = Threshold.

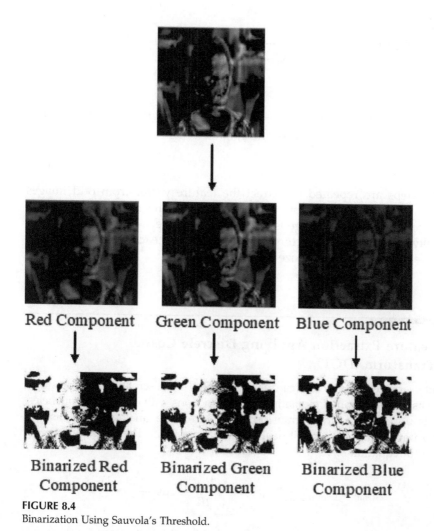

FIGURE 8.4
Binarization Using Sauvola's Threshold.

The above operations are repeated separately for all three channels of the image, namely, red (R), green (G) and blue (B). The pixel values assigned to 1 and 0 are clustered separately in two different clusters to calculate the mean of each cluster. Thus, the mean of the cluster with pixel values assigned to 1 is named the upper mean, and the mean of the cluster with pixel values assigned to 0 is named the lower mean as shown in equations 8.5 and 8.6. The upper and lower means thus calculated are considered the feature vector of the corresponding color component of the image. This action is repeated for all three color components of the images and thus the feature dimension becomes 6 for odd images.

$$Xupmean = \frac{1}{\sum\limits_{i=1}^{m} \sum\limits_{j=1}^{n} BitMap_x(i,j)} * \sum\limits_{i=1}^{m} \sum\limits_{j=1}^{n} BitMap_x(i,j)* X(i,j) \qquad (8.5)$$

$$Xlomean = \frac{1}{m*n - \sum\limits_{i=1}^{m} \sum\limits_{j=1}^{n} BitMap_x(i,j)} * \sum\limits_{i=1}^{m} \sum\limits_{j=1}^{n} (1 - BitMap_x(i,j))* X(i,j)$$

$$(8.6)$$

Similar steps are repeated to extract the feature vector from odd images without separating it into bit planes. Therefore, the number of features extracted from the odd image is also 6 on the whole.

Finally, the two feature vectors are associated with each other horizontally to create a feature vector of size 12.

8.4 Feature Extraction Applying Discrete Cosine Transform (DCT)

A finite series of data points can be represented by discrete cosine transform (DCT) as a summation of oscillating cosine functions with different frequencies. DCT is a separable linear transformation technique analogous to discrete Fourier transform (DFT), but it deals with real numbers. The working formula is given in equation 8.7.

$$X_{(m,n)} = a_m a_n \sum \sum R_{(i,j)} \; \cos \frac{\pi(2i+1)}{2R} \cos \frac{\pi(2j+1)}{2C} \qquad (8.7)$$

Rows and columns are denoted by R and C for the test image.

It is applied as a feature extractor on significant bit planes identified from the red, green and blue color components of the odd image variety created as the test image. The transformation technique is chosen for feature extraction due to its intrinsic property of shifting the high energy coefficients of an image toward the upper right corner and low energy coefficients toward the lower right corner. This property has enabled the selection of partial transform coefficients as the extracted feature vector which has drastically reduced the feature vector dimension. An illustration of this process is shown in Fig. 8.5.

As illustrated in Fig. 8.5, primarily DCT is applied on each of the color components to extract transform coefficients as feature vectors. Thus, the dimension of the feature vector during the initial phase is equal to the size of

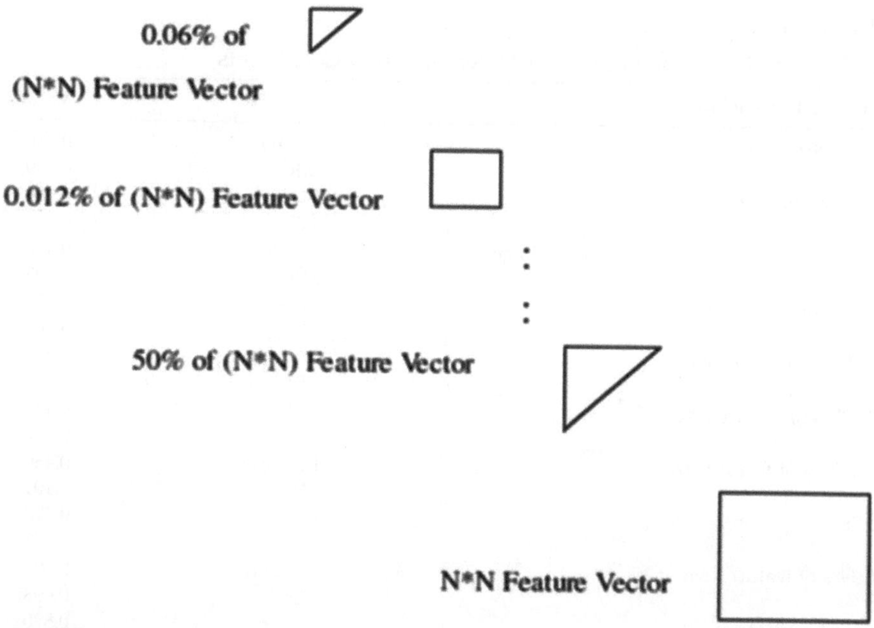

0.06% of

(N*N) Feature Vector

0.012% of (N*N) Feature Vector

.
.
.
.

50% of (N*N) Feature Vector

N*N Feature Vector

FIGURE 8.5
Extraction of Partial DCT Coefficients.

the test image. Furthermore, the dimension is reduced by capturing the higher energy coefficient in the upper left corner of the transformed image. Thus, in the second step, the feature vector is reduced to 50% and the process continues up to reduction of 0.06% of the original feature dimension. In each step of dimension reduction, the F1-Score and Misclassification Rate (MR) are measured to observe the classification results with the extracted feature dimension. An improving F1-Score and reduced MR are observed until 0.012%. Henceforth, the F1-Score decreases and the MR increases as shown in Table 8.1. Therefore, the selected fraction of transform coefficient for the feature vector is 0.012%, which has a dimension of 24.

8.5 Classification Framework

8.5.1 Method 1

Initially, classification results are evaluated separately for two different feature extraction techniques using a Euclidian Distance measure as in equation 8.8.

TABLE 8.1

F1-Score and MR Comparison for Different Partial Coefficients

Feature Dimension	Metrics	Values
100% feature size	F1-Score	0.230
	MR	0.169
50% feature size	F1-Score	0.360
	MR	0.135
25% of (N*N) feature size for N*N image	F1-Score	0.450
	MR	0.103
12.5% of feature size	F1-Score	0.450
	MR	0.103
6.25% of feature size	F1-Score	0.450
	MR	0.102
3.125% of feature size	F1-Score	0.450
	MR	0.102
1.5625% of feature size	F1-Score	0.460
	MR	0.101
0.7813% of feature size	F1-Score	0.470
	MR	0.101
0.39% of feature size	F1-Score	0.500
	MR	0.098
0.195% of feature size	F1-Score	0.520
	MR	0.095
0.097% of feature size	F1-Score	0.540
	MR	0.093
0.048% of feature size	F1-Score	0.580
	MR	0.087
0.024% of feature size	F1-Score	0.600
	MR	0.083
0.012% of feature size	F1-Score	0.630
	MR	0.078
0.006% of feature size	F1-Score	0.620
	MR	0.079

[**Observation:** An improving F1-Score and reduced MR are observed until 0.012% of partial transform coefficient as feature vector.]

$$D_{euclidian} = \sqrt{\sum_{i=1}^{n} (Q_i - D_i)^2} \qquad (8.8)$$

Henceforth, the classification process has been carried out using a late fusion framework as shown in Fig. 8.6.

The mentioned framework is based on decision fusion of two different distance measures, namely, Euclidian Distance and City Block Distance. The City Block Distance is given in equation 8.9.

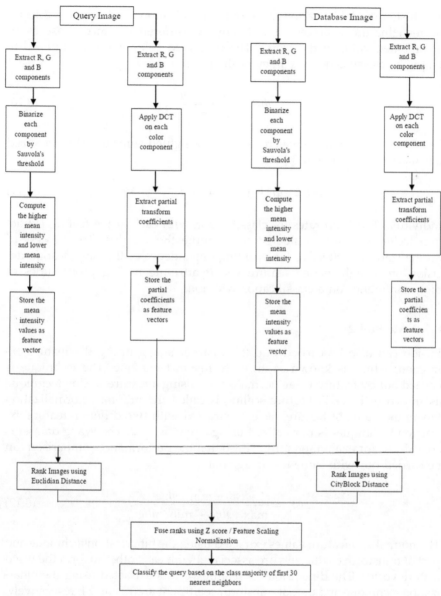

FIGURE 8.6
Late Fusion Framework.

$$D_{cityblock} = \sum_{i-1}^{n} |Q_i - D_i| \qquad (8.9)$$

The fusion model has applied z score normalization as in equation 8.10 for normalizing the distances computed by two different distance measures for a combined value of the scores. The combined value is further utilized to judge the classification accuracy of the technique.

$$dist_x = \frac{dist_x - \mu}{\sigma} \tag{8.10}$$

The final distance is calculated using the weighted sums of the normalized distances as in equation 8.11.

$$dist = w_1 d_n{}^{binarization} + w_2 d_n{}^{morphological} \tag{8.11}$$

Individual Precision rates for classification with each of the feature extraction techniques are considered to compute the weights. The process of normalization is effective for avoiding dependence of the classification decision on a high valued feature vector attribute having probabilities of higher influence on a classification decision.

8.5.2 Method 2

Fusion can also be carried out at an earlier stage during the feature extraction, which is known as an early fusion technique. The technique is carried out by feature value normalization using a feature scaling technique as shown in Fig. 8.7. Feature scaling is called the min-max normalization where the value of feature vectors extracted with two different feature extraction techniques is normalized using equation 8.12. The Wang dataset is considered for evaluation purposes so that the performance comparison can be carried out with same test image data.

$$dist_x = \frac{dist_x - \min\ value}{\max\ value - \min\ value} \tag{8.12}$$

The normalized feature values extracted using the binarization technique and partial transform coefficients are associated with each other to form the fused feature vector. The dimensions of feature vectors extracted using the binarization technique and partial transform coefficient are 12 and 24, respectively. Thus, the dimension of the fused feature vector becomes $(12 + 24) = 36$ per test image.

The fused feature is provided as input to four different classifiers, namely, K Nearest Neighbor (KNN), Random Forest (RF), Support Vector Machine (SVM) and multilayer perceptron (MLP).

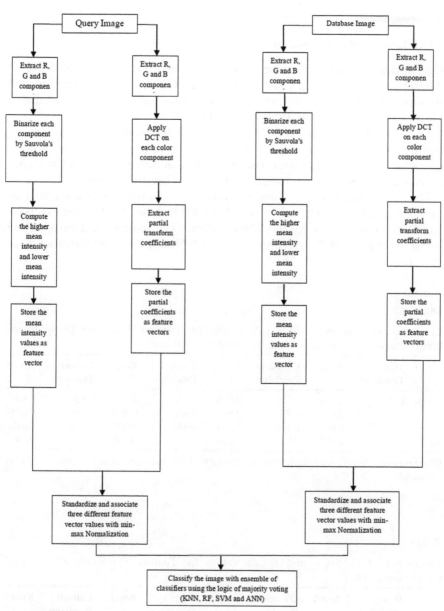

FIGURE 8.7
Early Fusion Framework.

8.6 Classification Results

Odd image varieties are generated for the test images in four different datasets, namely, Wang, Corel, Caltech and OT-Scene. The precision and recall values for each of the two feature extraction techniques are calculated individually as shown in Tables 8.2 and 8.3.

The classification results in Table 8.2 with four different classifiers for feature extraction using image binarization reveal superior classification using MLP. It is artificial neural network architecture and has shown higher Precision and Recall rates for all four datasets used for evaluation.

The classification results shown in Table 8.2 do not have the same pattern as those in Table 8.3. Unlike Table 8.2, the MLP shows the highest Precision and Recall values only for the Wang dataset. The remaining datasets show better Precision and Recall values for the KNN classifier than the other three classifiers.

TABLE 8.2

Evaluation of Precision and Recall Values for Feature Extraction Using Image Binarization with Sauvola's Local Threshold Selection

	Wang Precision	Recall	OT-Scene Precision	Recall	Corel Precision	Recall	Caltech Precision	Recall
KNN	0.710	0.670	0.590	0.560	0.500	0.450	0.520	0.490
RF	0.849	0.850	0.710	0.680	0.600	0.570	0.630	0.630
SVM	0.825	0.827	0.680	0.690	0.580	0.560	0.610	0.610
MLP	0.866	0.867	0.710	0.720	0.610	0.580	0.630	0.640

[**Observation:** Classification with MLP has the highest Precision and Recall values for all four datasets.]

TABLE 8.3

Evaluation of Precision and Recall Values for Feature Extraction with Partial Coefficient of DCT

	Wang Precision	Recall	OT-Scene Precision	Recall	Corel Precision	Recall	Caltech Precision	Recall
KNN	0.825	0.818	0.916	0.913	0.841	0.82	0.851	0.837
RF	0.820	0.819	0.828	0.827	0.682	0.681	0.787	0.786
SVM	0.858	0.851	0.619	0.622	0.410	0.420	0.466	0.464
MLP	0.841	0.841	0.743	0.749	0.434	0.431	0.464	0.465

[**Observation:** Highest Precision and Recall values are observed for all four classifiers with different datasets.]

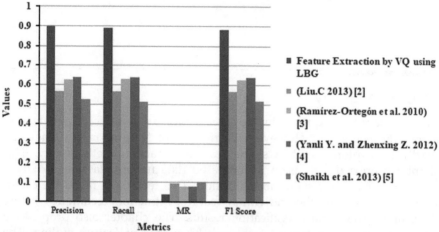

FIGURE 8.8

Comparison of Precision and Recall Values of Fusion-Based Classification Techniques. [**Observation:** The highest Precision and Recall values are observed with the late fusion technique compared to the early fusion and single feature vector extraction techniques.]

Thereafter, the two fusion-based classification techniques, namely, the early fusion (feature fusion) and late fusion (classification decision fusion) techniques are carried out to evaluate classification accuracy. The experiment is performed with the Wang dataset.

The results in Fig. 8.8 clearly establish that the late fusion technique (decision fusion-based classification) has better Precision and Recall values than the early fusion technique (feature fusion-based) and has improved classification accuracy.

However, both the fusion-based techniques for classification exhibit superior performance compared to the individual techniques of classification in Tables 8.2 and 8.3 carried out with a single feature vector of the test image.

Fig. 8.9 reveals higher Precision and Recall values for the late fusion technique compared to the benchmarked techniques.

Therefore, the fusion-based technique appears to be highly efficient in ensuring high classification accuracy. However, the extracted feature vector dimensions involved in the process of fusion must be small to avoid elevated convergence of time and high computational overhead.

Thus, extraction of a robust feature is not the only parameter for enhanced classification accuracy. In this chapter, two feature extraction techniques are used in which the feature dimensions are 12 and 24, respectively. The small feature size crafted for both techniques kindled the opportunity to adopt fusion-based techniques for content-based image identification. Effective feature dimension reduction that retains the essential feature components is inevitable to ensure efficient content-based image classification with a high level of accuracy.

Chapter Summary

Conventional techniques of image classification are based on single feature extraction techniques. However, the rich set of features in an image can hardly be explored with a single feature extraction methodology. This chapter proposed a decision fusion-based (late fusion) classification technique by means of z score normalization. This technique fused the classification decision of the individual feature extraction techniques before determining the final classification results. The chapter also proposed a feature fusion (early fusion) technique by means of feature scaling. The introduced fusion techniques outclassed the classification results obtained using the single feature extraction technique. Comparing the fusion techniques resulted in higher classification accuracy for late fusion using classification decision.

References

1. Thepade, S., Das, R. and Ghosh, S., 2017. Decision fusion-based approach for content-based image classification. *International Journal of Intelligent Computing and Cybernetics*, 10(3): 310–331.
2. Das, R., Thepade, S. and Ghosh, S., 2016. Framework for content-based image identification with standardized multiview features. *Etri Journal*, 38(1): 174–184.

9

Future Directions: A Journey from Handcrafted Techniques to Representation Learning

9.1 Prelude

Preceding years have witnessed the widespread popularity of handcrafted techniques in defining image descriptors for content-based image classification. Several low-level features including shape, color, texture, etc. are utilized to design descriptors from the content of image data. These descriptors, popularly known as feature vectors of images, are utilized as inputs to a classifier for building the classification model for content-based image classification.

Nevertheless, recent times have embraced the power of artificial neural networks in almost every field of information extraction for any kind of data. The architecture of the network has proven to be significantly robust in extracting meaningful insights from multiple types of data without any human intervention. Convolutional neural networks (CNNs) have gained immense popularity for use in extracting content-based features from image data [1,2]. The reason behind such wide acceptance is the availability of CNNs that are already trained over a huge dataset with extensive class varieties. Training of CNNs on large datasets such as ImageNet has made the architecture capable of extracting credible features from similar datasets. Some popular pretrained CNN architectures include VGG, ResNet, and MobileNet.

Representation learning is based on the hierarchical feature learning property of deep learning models. A convolutional neural network is a deep learning network comprised of a convolutional base and a classifier, as shown in Fig. 9.1. The convolutional base is responsible for feature computation, which is generic in nature and can be used for diverse problem areas. However, the last layer features are dataset-specific and are used for a particular type of problem. Robust feature representation is performed automatically by the layers of the pretrained convolutional neural network, which has eradicated the need for a manual, domain-specific algorithm

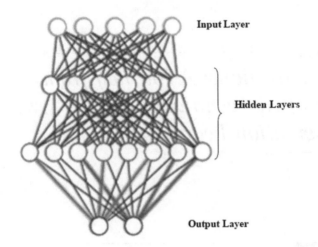

FIGURE 9.1
Structure of Convolutional Neural Networks.

design. Features for any target dataset can be extracted from any layer of choice of the pretrained CNN, which makes the model immensely flexible.

Hence, the automated feature extraction technique in pretrained CNNs is gaining broad acceptance and gradually making the current state-of-the-art handcrafted techniques irrelevant.

9.2 Representation Learning-Based Feature Extraction

Feature extraction using representation learning is considered one of the most essential characteristics of CNNs. Numerous benchmarked hand-crafted techniques have been developed for feature extraction. Some of the most prominent techniques are local binary pattern (LBP), SIFT, SURF, and histogram of oriented gradients (HoG). However, all of these techniques have proven to be either too generic or overengineered. This has stimulated the popularity of representation learning through use of a filter. The filter is designed with a user-specified dimension and is shifted across a given image in a direction from top left to bottom right.

The filter helps to determine the frequency, count and location of a particular feature in an image. The convolutional operation of a filter is shown in Fig. 9.2.

Implementation of representation learning is shown in following section.

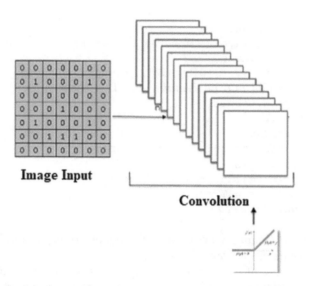

FIGURE 9.2
Convolutional Operation.

9.3 Code Example (MATLAB®)

#Loading Image Dataset
```
unzip('MerchData.zip');
imds=imageDatastore
('MerchData','IncludeSubfolders',true,'LabelSource',
'foldernames');
[imdsTrain,imdsTest]=splitEachLabel(imds,0.7,'randomized');
numTrainImages=numel(imdsTrain.Labels);
idx=randperm(numTrainImages,16);
figure
for i=1:16
subplot(4,4,i)
I=readimage(imdsTrain,idx(i));
imshow(I)
end
```
#Load Pretrained Network
```
net=resnet18
inputSize=net.Layers(1).InputSize;
analyzeNetwork(net)
```
#Extract Image Features
```
augimdsTrain=augmentedImageDatastore(inputSize
(1:2),imdsTrain);
augimdsTest=augmentedImageDatastore(inputSize(1:2),imdsTest);
layer='pool5';
```

```
featuresTrain=activations
(net,augimdsTrain,layer,'OutputAs','rows');
featuresTest=activations
(net,augimdsTest,layer,'OutputAs','rows');
```
   #*Extract Class Labels from Training and Test Data*
```
YTrain=imdsTrain.Labels;
YTest=imdsTest.Labels;
```
   #*Fit SVM Classifier*
```
classifier=fitcecoc(featuresTrain,YTrain);
```
   #*Classify Test Images*
```
YPred=predict(classifier,featuresTest);
accuracy=mean(YPred==YTest)
```
   #*Classification Accuracy*
```
accuracy=1
```
   Source: https://in.mathworks.com/help/deeplearning/examples/extract-image-features-using-pretrained-network.html

A brief overview of content-based image classification techniques and a comparison of their performance across domains is presented in the following sections.

9.4 Image Color Averaging Techniques

The techniques of sorted block truncation coding (SBTC) and block truncation coding (BTC) with color clumps have been compared to evaluate their performance. The results clearly indicate the supremacy of feature extraction using BTC with color clumps over the SBTC method as well as state-of-the-art techniques of feature extraction in classification performance.

Image averaging techniques average the color content of images, which can be considered a feature vector for content-based image classification. Use of color averaging techniques for feature extraction help obtain faster and better image classification. SBTC and BTC with color clumps are implemented for feature vector extraction from images.

SBTC divides intensity values into N blocks where ($N>=2$). Feature extraction is carried out following the BTC technique, where each color component of red (R), green (G) or blue (B) is considered a block. Each block is divided into bins of sorted intensity values. The average of the sorted intensity values in each bin forms the feature vector of that block.

Another feature extraction technique is BTC with color clumps. The color clumps approximate the distribution of gray values in each color component and create disjoint subsets of gray values for each color component. This color clumps method logically divides the dimension of the intensity values in each color component of red (R), green (G) or blue (B) into a different number of partitions at each stage of feature vector extraction. The mean threshold for each color clump is calculated, and each gray value is compared to the threshold. The values higher than the threshold are clustered in the upper intensity group, and the values lower than the threshold are clustered in the lower intensity group. The mean of the two clusters thus formed is considered the mean upper intensity and mean lower intensity for the corresponding clump. Hence, the number of feature vectors is twice the number of color clumps formed and is independent of the size of the image.

9.5 Binarization Techniques

Two different approaches of threshold selection can be implemented to extract feature vectors based on binarization techniques: mean threshold selection and local threshold selection.

Binarization with mean threshold selection computes a threshold value for each color component red (R), green (G) or blue (B) in an image. Binary bitmaps are then calculated for each color component in RGB color space. A value of 1 is assigned to pixel values greater than or equal to the threshold value; otherwise, 0 is assigned. Two mean colors are derived for each color component after calculating the bitmap values. The upper mean color is the average of the pixel values that are greater than or equal to the consequent threshold, whereas the lower mean color is derived from the average of the pixel values that are lower than the threshold. Four different feature extraction techniques by binarization using mean threshold selection are proposed in Chapter 4, which are feature extraction with multilevel mean threshold, feature extraction from significant bit planes using mean threshold selection, feature extraction from even and odd image varieties using mean threshold selection and feature extraction using static and dynamic ternary image maps using mean threshold selection.

Feature extraction with local threshold selection for binarization is carried out using either Niblack's local threshold selection or Sauvola's local threshold selection. A local window of dimension $m*n$ is maintained for the binarization process carried out with local threshold selection. Pixel-wise threshold is calculated by the method for each color component by sliding a rectangular window over the component. The image may have uneven illumination for which the binarization process is carried out locally using the rectangular window.

The feature extraction techniques using binarization with mean threshold selection and local threshold selection have been compared for classification performance. Feature extraction by binarization with Sauvola's local threshold selection has outclassed all the other techniques in Precision, Recall and F1-Score and has shown minimum Misclassification Rate (MR). This technique has also surpassed the existing techniques of feature extraction in classification performance.

9.6 Image Transforms

The five transform techniques that have been proposed for feature extraction are discrete cosine transform (DCT), Walsh transform, Kekre transform, discrete sine transform (DST) and discrete Hartley transform (DHT). A comparison of classification performance for these techniques shows that feature extraction with partial DCT coefficients has outclassed Walsh transform, Kekre transform, DST, and DHT. The supremacy of the proposed technique has also been established over the existing techniques for classification performance.

The purpose of any image transform is to relocate the high frequency components toward the upper end of the image and the low frequency components toward the lower end of the image. Transformation drastically reduces the size of feature vectors by excluding insignificant coefficients. Energy compaction of transforms can pack a large fraction of the average energy into a few components. This has led to faster execution and efficient algorithm design for content-based image classification.

The extracted feature vectors from each component in transform coefficients are stored as a complete set of feature vectors. In addition, partial coefficients from the entire feature vector set are extracted to form the feature vector database. Feature vector databases with partial coefficients are constructed, starting with 50% of the complete set of feature vectors to 0.06% of the complete set of feature vectors.

9.7 Morphological Operations

Feature extractions with two different morphological operators, called top-hat and bottom-hat morphological operators, are compared for classification performance evaluation.

Shape is considered a vital feature for identification of object-based information. Shape features play a vital role in image recognition by identifying the structure of the object of interest with the help of structuring elements. Morphology-based operation is implemented for feature extraction from images using top-hat and bottom-hat operators. Basic morphological operations are performed by dilation and erosion functions. Initially, the image is processed with swelling by applying pixels to the object boundaries with dilation; the image is shrunk with erosion by removing the boundary pixels. The corresponding pixel in the input image is compared with the neighboring pixels to define the value of each pixel in the output image.

The contour region and background portion of individual color components of the image are grouped into two different clusters. The mean and standard deviation for the two clusters are derived by computing the feature vectors of the images.

A comparison of feature extraction using top-hat and bottom-hat morphological operators has clearly revealed that feature extraction with bottom-hat transform has better Precision, Recall and F1-Score values and a lower MR. A comparison with state-of-the-art techniques has yielded the same results.

9.8 Texture Analysis

Two techniques of feature extraction using texture analysis, called feature extraction by vector quantization codebook representation using Linde-Buzo-Grey (LBG) algorithm and feature extraction by gray level co-occurrence matrix (GLCM), have been compared for classification performance.

In some images, the defining characteristics of regions can be important for proper analysis. Feature extraction can be done from the spatial arrangements of color or intensities with the help of texture analysis. Some histogram distributions can have different texture representation, which can act as a tool for extraction of distinct features. Vector quantization (VQ) is used to generate codebook as feature vectors from the images. A k dimensional Euclidian space is mapped by means of vector quantization into a finite subset. The codebook is represented by the finite set CB.

Spatial definition of visual pattern of images has been characterized by the texture features. Specific texture detection in an image is performed by representing the texture as a two-dimensional gray level variation known as the gray level co-occurrence matrix (GLCM). GLCM is a statistical method for investigating textures that considers the spatial relationship of pixels. The texture of an image is portrayed by the GLCM function by computing the frequency of existence of pixel pairs with specific values and specific pixel relationships in an image followed by extraction of statistical measures from the matrix.

The comparisons based on Precision, Recall, MR and F1-Score have established that feature extraction with VQ using LBG is superior to the GLCM method of feature extraction for content-based image classification and to the state-of-the-art techniques.

9.9 Multitechnique Feature Extraction for Decision Fusion-Based Classification

The competence of the content-based image identification technique depends on the extraction of robust feature vectors. Diverse low-level features such as color, shape, texture, etc. constitute the process of feature extraction. However, an image comprised of a number of features can hardly be defined by a single feature extraction technique. Therefore, two different fusion-based classifications are carried out and compared with one another. The first technique has carried out late fusion by separately evaluating classification decision of two different feature vectors, namely, binarization features extracted with Sauvola's local threshold selection and transform domain features extracted with partial coefficients of Discrete Cosine Transform (DCT). The second fusion based technique is early fusion of the above mentioned feature vectors. The binarization feature and transform domain feature are normalized separately using z score normalization followed by horizontal concatenation of the two. The combined feature vector with larger details of image data is now provided as an input ti the classifier for better performance.

The two results for classification with fusion techniques are compared, and the results indicate that classification with decision fusion of four feature extraction techniques surpasses classification with decision fusion of three different feature extraction techniques.

Furthermore, the superior technique is used as a precursor for retrieval, and the query image is classified by the proposed fusion-based approach. The retrieval results are compared to the state-of-the-art techniques of fusion-based retrieval, and the proposed method outperforms the existing fusion-based techniques.

9.10 Comparison of Cross Domain Feature Extraction Techniques

The feature extraction techniques from binarization, image transform, morphological and texture analysis are compared for classification performance. The best technique selected in each domain is used for comparison in Fig. 9.3.

The comparison in Fig. 9.3 clearly reveals that classification with feature extraction by VQ using LBG has the best Precision, Recall, F1-Score and MR compared to the rest of the proposed feature extraction techniques in other domains.

9.11 Future Work

The book extensively covers content-based feature extraction techniques in assorted domains for image classification. The classification results divulge statistical significance of improved performances over state-of-the-art

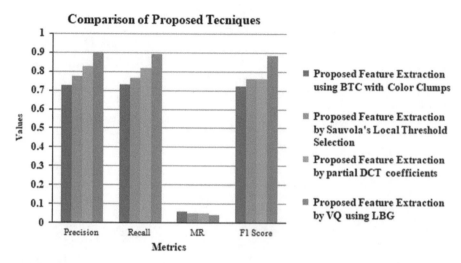

FIGURE 9.3
Comparison of Classification with Cross Domain Feature Extraction Techniques.
[**Observation:** Classification by feature extraction with VQ using LBG has the best performance.]

techniques. Henceforth, these techniques can be evaluated for image datasets in diverse domain including medical imaging, satellite imagery, plant disease detection, and so on. The fusion methodologies may also be implemented for combining representation learning features to handcrafted descriptors to assess the classification performances of the hybrid feature definition.

The book provides a novel perspective to computer vision aspirants while they handle image data for classification. Moreover, the fusion-based propositions may surpass the accuracy of neural networks when used to combine traditional features with representation learning-based features.

References

1. Sainath, T. N., Mohamed, A. R., Kingsbury, B. and Ramabhadran, B., 2013, May. Deep convolutional neural networks for LVCSR. In 2013 IEEE International Conference on Acoustics, Speech and Signal Processing, pp. 8614–8618.
2. Zhang, Y. D., Dong, Z., Chen, X., Jia, W., Du, S., Muhammad, K. and Wang, S.H., 2019. Image based fruit category classification by 13-layer deep convolutional neural network and data augmentation. *Multimedia Tools and Applications*, 78(3): 3613–3632.

10

WEKA: Beginners' Tutorial

10.1 Prelude

Waikato Environment for Knowledge Analysis (WEKA) is an open source data mining tool developed by the University of Waikato in New Zealand [1]. This book has extensively used WEKA to evaluate classification results of content-based image data.

WEKA is considered a technique for machine learning using graphical user interface (GUI). The software has a vast collection of machine learning algorithms that are capable of data preprocessing, classification, regression, clustering, association rules and visualization [2].

10.2 Getting Started with WEKA

Please use the following snapshots to download WEKA:

1. The search results for 'download Weka' phrase is shown (Fig. 10.1). Click on the link provided below to download. https://www.cs. waikato.ac.nz/ml/weka/

2. Click on the DOWNLOAD AND INSTALL tab (Fig. 10.2).

3. Select the option for your installed operating system (OS) and download the self-extracting files by clicking the appropriate choice (Fig. 10.3).

4. Install WEKA (Fig. 10.4).

5. Initiate the WEKA application (Fig. 10.5).

6. Click the Explorer tab on WEKA GUI Chooser (Fig. 10.6).

7. Click the Open File tab in WEKA Explorer (Fig. 10.7).

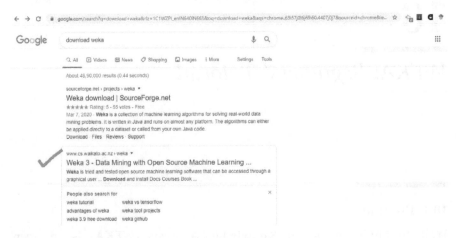

FIGURE 10.1
Google Search for WEKA.

8. Select a data file having any of the file extensions shown in the drop-down menu (Fig. 10.8).

9. Click the Classify tab followed by the Choose tab (Fig. 10.9).

10. Select Classifier from the drop-down menu (Fig. 10.10).

11. Click on the Start tab and view Results (Fig. 10.11).

FIGURE 10.2
WEKA Home Page.

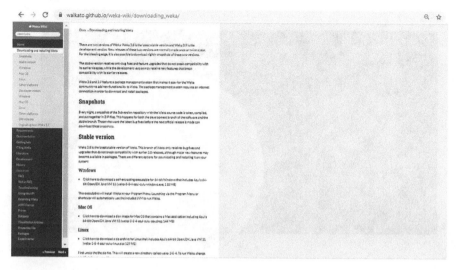

FIGURE 10.3
Download WEKA.

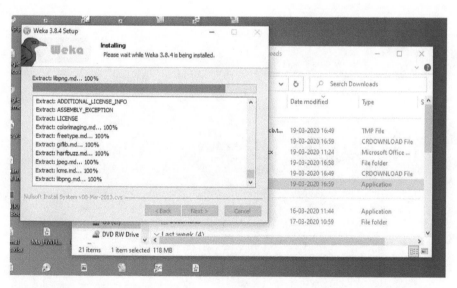

FIGURE 10.4
WEKA Installation.

FIGURE 10.5
Initiate WEKA.

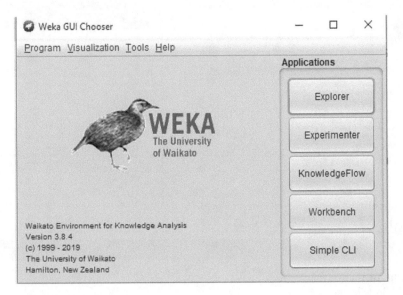

FIGURE 10.6
WEKA Home Screen.

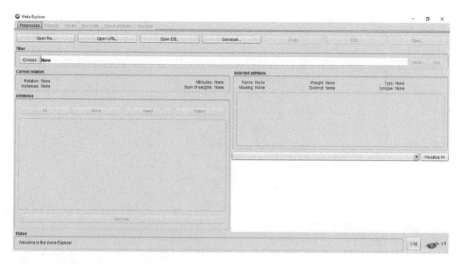

FIGURE 10.7
Select Data File.

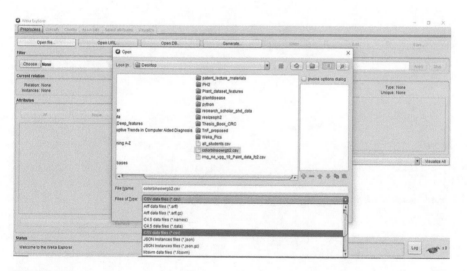

FIGURE 10.8
Upload Data.

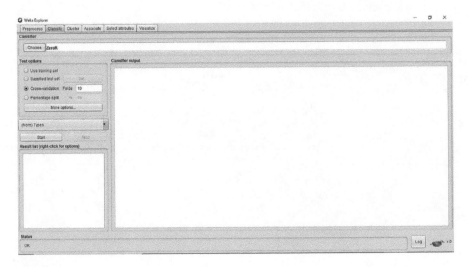

FIGURE 10.9
Choose Classifier.

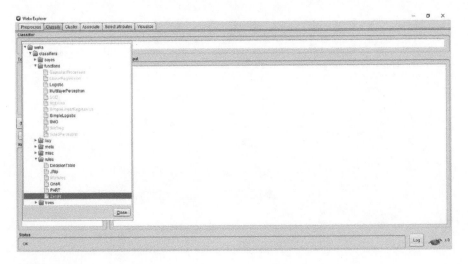

FIGURE 10.10
Select Classifier.

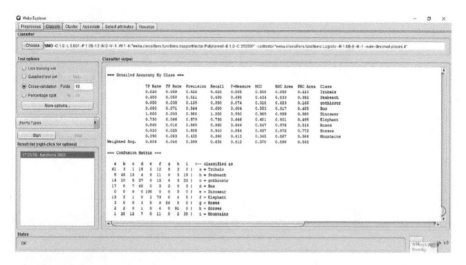

FIGURE 10.11
WEKA Classification Results.

References

1. Hall, M., Frank, E., Holmes, G., Pfahringer, B., Reutemann, P. and Witten, I. H., 2009. The WEKA data mining software: an update. *ACM SIGKDD explorations newsletter*, 11(1): 10–18.
2. Frank, E., Hall, M., Trigg, L., Holmes, G., and Witten, I. H., 2004. Data mining in bioinformatics using Weka. *Bioinformatics*, 20(15): 2479–2481.

Index